Mixing in Flow

Mixing in Flow is a focused guide for mix engineers who are familiar with the basics of mixing and now seek greater consistency and refinement in their work. Rather than covering every fundamental aspect of mixing, it picks up where many mixers experience a plateau – when they know the mechanics but struggle with inconsistent or lacklustre results.

This book begins by defining the art of mixing, before moving into an intensive bootcamp where readers delve deep into working with the essential tools and developing effective workflows to maximise their potential. Within this framework, special emphasis is placed on ear training and critical engagement with sound. Using a problem-based approach, the focus then shifts to mixing in practice, by exploring common mixing challenges from the perspective of what we hear, with various techniques offered for addressing these sonic issues. The final section covers advanced topics, providing further insights that follow on from the development of core mixing skills.

Designed to inspire readers to develop their unique mixing flow and have confidence in their tools, *Mixing in Flow* is the ideal companion for anyone looking to improve their mixing, including students and aspiring professional engineers, as well as more experienced mixers looking for new inspiration and continued professional development.

Jane Arnison is a multi-disciplinary sound professional with a career spanning popular and experimental music. Her expertise combines a conservatoire-based music bachelor's degree, a degree in philosophy and a master's in sound art. Jane's creative work reflects her diverse background, ranging from collaborations with major labels like Sony Music, mentoring emerging talent, and collaborating on large-scale immersive installations. As a lecturer in production engineering and sound art, Jane teaches at NYU's Berlin Campus and at FHNW Basel's Jazz Campus. She frequently leads workshops and speaks at conferences and festivals on sound.

Mixing in Flow

A Critical Approach to Mixing Sound

Jane Arnison

LONDON AND NEW YORK

Designed cover image: watchara_tongnoi via Getty Images

First published 2025
by Routledge
4 Park Square, Milton Park, Abingdon, Oxon OX14 4RN

and by Routledge
605 Third Avenue, New York, NY 10158

Routledge is an imprint of the Taylor & Francis Group, an informa business

© 2025 Jane Arnison

The right of Jane Arnison to be identified as author of this work has been asserted in accordance with sections 77 and 78 of the Copyright, Designs and Patents Act 1988.

All rights reserved. No part of this book may be reprinted or reproduced or utilised in any form or by any electronic, mechanical, or other means, now known or hereafter invented, including photocopying and recording, or in any information storage or retrieval system, without permission in writing from the publishers.

Trademark notice: Product or corporate names may be trademarks or registered trademarks, and are used only for identification and explanation without intent to infringe.

British Library Cataloguing-in-Publication Data
A catalogue record for this book is available from the British Library

ISBN: 978-1-032-26646-6 (hbk)
ISBN: 978-1-032-26644-2 (pbk)
ISBN: 978-1-003-28923-4 (ebk)

DOI: 10.4324/9781003289234

Typeset in Sabon
by codeMantra

Access the Support Material: https://www.janearnison.com/mixing-in-flow-online-resources

Contents

List of figures	*xv*
Acknowledgements	*xvii*
Introduction to the book	*xix*
About the author	*xxi*

PART ONE
Get it right from the start   1

1   What is mixing?   2
     Mixing is not production 2
     A working definition for mixing 3
     Critical listening with a mixing mindset 3
     Decision making – from listening a plan emerges 5
     Targeted technical processing 5
     Optimising the aesthetic identity 6
     Revision of this section and moving on to Part Two 6

PART TWO
Mixing bootcamp – getting the most out of our tools and ourselves   9

2   Interacting with the multitude of tools   10
     Equalisation 10
       Different equaliser circuits 11
       What about digital equalisers? 13
       Transformers colouring sound 13
       Controls 14
       AI and adaptive equalisation 16
       Phase and equalisation 16
     Compression 18
       More than just level balancing 18
       Various types of compressors 18

One compressor to rule them all? 21
Uncovering the secrets of our tools 21
The magic of manuals 23
Other dynamics processing 24
　Gates and expanders 24
　Dynamic equalisation vs multiband compression 24
　Sidechaining 25
Saturation and distortion 26
　Some definitions 26
　Other artefacts that may occur 27
　Smearing the transient 28
　Frequency changes 28
　Specific mechanical artefacts 28
Phase 30
　What is phase? 30
　What are phase issues? 30
　What are phase processes? 31
　The magic of phase 31
　Psychoacoustics and the precedence effect 31
　What is the echo threshold? 32
Space processes 34
　Time and space converge! 34
　Analog vs algorithmic vs convolution/impulse response vs AI 34
　Connecting to the aesthetic of the production 35
　Different perspectives and uses 35
　Old school still cool 35
　Parallel or insert? 35
　Adjusting and tuning 36
　How many is too many? 36
Meters 38
　Digital meters 38
　Peak meters 38
　RMS meters 39
　What levels should I set my mix to? 40
　Analog Meters 40
　New(er) formats 42
　… a parting thought 43

3　**From theory to practice**　　　　　　　　　　　　　　　　　　　45
Step 1: building your mixing toolkit 45
　The mixing tool kit 46
Step 2: learn your tools 48
Step 3: gaining the confidence to know which tool to use when 49
Step 4: expanding or adjusting the tool kit 49
　Expanding your tool kit with an analog approach 49

POP quiz: am I a master of my tools? 50
   Equalisation 50
   Compression, dynamics and envelope shaping 50
   Saturation 50
   Phase processes 51
   Space 51
   Metering 51

4 Developing advanced ear sensitivity   52
Note on practice 52
Exercise explanation 52
   Amplitude recognition 53
   Compression, saturation, clipping 53
   Space and phase 53
Ear training boot camp 53
   Beginner 54
   Intermediate exercises 54
   Advanced exercises 55

5 Setting up your listening environment   56
Monitoring 57
   Stick with your monitoring 57
   Can I or should I use headphones? 57
The room 58
   Room analysis tools 58
Making alterations to your room 58
   Bedroom mixing 58
   Commencing treatment in a multi-purpose space 59
   Doing self-treatment in your devoted mix studio 60

6 The power of references   62
A vital tool for mix engineers 62
   Why don't aspiring engineers use them? 62
Building a reference playlist 63
Uses for references when mixing 63
   Exercise: mixing playlist 64
Setting up references when mixing 64
   Loudness matching 64
   Echoic memory – the importance of the quick switch 65

7 Workflow   66
Mixing in flow 66
   How to get into flow states when mixing? 66
Embodiment 68
   How does embodiment relate to mixing? 69

How do I become embodied? 69
Establishing a workflow 70
  A guide to approaching a mix that you actually finish 70

8 The pre-mix preparation   72
Set it up right and make life easy 72
  DAW or desk set up 72
  Set the sample rate and bit depth 72
  Import stems organise name colour 72
  Set up your meters 72
Gain staging 73
Routing 74
Set up your references 75
Setting up the mix bus 75
Note on workflow: direction through the mix 76
How do you know when you are ready to begin mixing? 77

PART THREE
Creative problem solving – inside the mix   79

9 Not everything can be front and centre   80
What are we hearing when we are listening? 80
Concepts discussed in this section 80
Organising and prioritising 80
Building a mix picture 81
The dimensions of listening 81
Setting initial balances 81
Playing with our perceptions 82
  Width 82
  Depth 82
  Height 82
Time, perception and relativity! 83

10 Clean it up   84
What are we hearing when we are listening? 84
Concepts discussed in this section 84
Frequency 84
  Optimising individual sounds 84
  To solo or not to solo 84
  To sweep or not to sweep 85
  Accountable EQ or turn it off 86

    What type of equaliser should I use? 87
    Clicks and pops 87
        Where do they come from? 87
        How do I fix them? 87

11 Don't get stuck in the mud   88
    What are we hearing when we are listening? 88
    Concepts discussed in this section 88
    Frequency, register and arrangement 88
    Frequency masking 90
    Repositioning in space 91
    Side chaining 92

12 Fighting for attention   93
    What are we hearing when we are listening? 93
    Concepts covered in this section 93
    Frequency 94
    Transient 94
        How do we move the transient of one sound out of the way? 94
    Envelopes and shapes 95
        What tools can I use to adjust envelopes? 95
    Further technical solutions 96
        Side chaining 97
        Multi-band compression or dynamic EQ 97

13 More power!   98
    What are we hearing when we are listening? 98
    Concepts covered in this section 98
    Phase issues 98
    Arrangement problems 99
    Frequency masking 99
    Dynamic processing 100
    Envelope 100
    Saturation 100

14 Thicken up the sauce   101
    What are we hearing when we are listening? 101
    Concepts we will cover in this section 101
    Using saturation to create thickness 101
        Individual sounds 101
        Groups of sounds 102
    Using space to create thickness 102

15 Let's get this party started · 104
What are we hearing when we are listening? 104
Concepts we will cover in this section 104
Getting into the groove 104
Envelopes and envelop shaping 104
  Compression as a method for envelope shaping 105
  Gating and downward expanding 105
  Connecting elements to each other through side chaining and envelop following 105

16 Low-end theory · 107
What are we hearing when we are listening? 107
Concepts we will cover in this section 107
Challenges that block our efforts 107
Relationship between the parts 108
Don't get stuck in the mud 109
Identifying phase issues 110
Dynamics and envelope 110
Saturation 111

17 I feel it all around me · 112
What are we hearing when we are listening? 112
Concepts covered in this section 112
Aligning spaces with the production aesthetic 112
Relationship between space and emotion 113
Perceptions of mix depth and width 113
Tips for working with space processing 113
Tuning the spaces 114

18 They only listen to the vocals anyway · 115
The voice is a particularly challenging element for modern mixing for several reasons 115
Lead vocal 116
  What are we hearing when we are listening? 116
  Concepts covered in this section 116
  Clean it up 116
  Dynamics 118
  Add extra sparkle 120
  Make the vocal larger than life 121
Backing vocals 122
  What are we hearing when we are listening? 123
  Concepts we will cover in this section 123
  Clean it up and bounce it 123
  Optimising the layers, clearing space 123
  Panning approaches 124

Contents xi

    Create cohesion 124
    Keep the BV's out of the way of the lead! 124
  Group vs individual processing 125
  Pitching and timing 125
    Some tips on pitch correction 126

19 Take me on a journey!   128
  What are we hearing when we are listening? 128
  Concepts discussed in this section 128
  Don't start too high 128
  Composing the joins 129
  Golden mean 129
  Subtle changes and automations 129
  Don't be too perfect! 129

PART FOUR
Important next considerations   131

20 Preparing for a mix – moving from production to mixing   132
  Mixing your own work 132

21 The role of a mix engineer in the collaborative process   134

22 Psychoacoustics   136
  What is psychoacoustics? 136
  How do psychoacoustic concepts relate to mixing? 136
    Noise exposure and hearing sensitivity 136
    The acoustic reflex – dulling effect of sounds after exposure 136
    Loudness perception 137
    Just noticeable difference 137
    Critical bands 138
    Precedence effect and Haas effect 138
    Directional illusion 138
    Mixing is psychoacoustics! 139

23 Aesthetics and individual mixing style   140
  What are mix aesthetics? How are they different from the overall
    sonic aesthetic? Is a mix aesthetic different from production aesthetic? 140
  Saying goodbye to references? 141

24 Mixing whilst producing   142
  The argument for separating mixing and producing 142
  The argument for combing mixing and producing 143
  Some considerations when mixing whilst producing 143
  Mixing in a genre bending landscape: towards experimentalism 144

25 Mix bus processing — 146
Evolution of mix bus processing 146
Top-down mixing 147
How to set up your mix bus 147
Good practice tips 147

26 Loudness — 148
The loudness wars 148
Genre and loudness 148
Loudness starts before the mix 149
Loudness optimisation during mixing 149
Loudness and the mix bus/master 149
Loudness and mastering for a mixer 150

27 Impact of AI on mixing — 152
Types of AI tools 152
AI as a co-creative tool 153
Future prediction 153
Final word, for now… 153

28 Spatial mixing – mixing beyond stereo — 155
What is spatial mixing? 155
Is it a fad? 155
The rise and fall and rise of spatial audio 156
Speaker layout 158
Different approaches to spatial mixing 160
  Channel-based multichannel (e.g. 5.1) 160
  Object based 161
  Dolby Atmos 162
  Ambisonic 163
  Advice for choosing which approach to adopt 164
Key concepts of spatial mixing 165
  Monitoring 165
  Binaural mixing/monitoring (headphones) 165
  Panning 166
  Spatialising 167
  Encoding/decoding 167
  Rendering 167
  Routing/bussing 168
  LFE 168
Best practice tips 168
  Prep the audio 169
  Setting up and routing 169

The key processes of mixing are still frequency and dynamics 170
Low frequency management 170
Loudness 170
Depth and space 171
Spatialisation – placement 171
Spatialisation – movement vs stability 172
Workflow 172

Final word   174
Rules are meant to be broken 174
The learning never ends 174

Bibliography and expanded reading list   *177*
Index   *179*

Figures

1.1	Mixing components and relationships	3
2.1	Analog mixing console	11
2.2	Transformers in the back of audio hardware	14
2.3	Different EQ slopes	14
2.4	API and 1073 Equalisers	15
2.5	Various types of compressors	19
2.6	Optical circuit	20
2.7	Clipping wave shaping diagrams	26
2.8	Images showing in in and out of phase	30
2.9	Standard meters in various DAW	38
2.10	Metering scale comparing analog and digital	41
2.11	VU meter	41
2.12	Correlation between VU and other meters	41
5.1	First reflection path	60
8.1	VU meter on the master bus showing the dial at 0 for the most part of the mix means gain staging is complete	73
8.2	Routing example for a typical mix	75
11.1	Frequency range diagram including instrumental registers	89
11.2	Characteristics of sound over frequency range	91
12.1	ADSR diagram	95
12.2	Visual representation of sounds	95
18.1	Example of a lead vocal chain	121
22.1	Fletcher Munson's curve	137
28.1	Different approaches to speaker placement	158
28.2	Surround panner example	160
28.3	Routing in Dolby Atmos	163
28.4	Routing in ambisonic	164
28.5	Mono vs stereo localisation diffusion	172

Acknowledgements

I want to start by expressing my gratitude to my wife, Julie. She supports me and encourages me with all my endeavours. She is always there to offer advice and also to remind me to slow down when I need to hear it. In particular her encouragement when I'm full of doubt always lifts me up.

I would also like to thank the generous words and support from my idols and mentors who I have met along the way, including the legendary Russell Elevado and Bob Power. Their mixing inspired me and ultimately planted the seed for my journey down this path. Additionally, I am grateful for those who have supported my efforts to expand opportunities in the industry, particularly Erik Brewer from Brewery Studios in Berlin. A special thanks goes to my friend and mentor Adam Townsend from APRA, who has consistently provided me with energy and support to take on new challenges and offers me numerous opportunities. I also want to thank Dr. Julia Schröder from the UDK, who helped me begin my writing journey by supervising the first draft of the initial section of this book. Without her structural support, I might not have taken the first step.

I would also like to thank the amazing team at New York University Berlin and Clive Davis Institute in New York for their support and the Global Faculty Fund for awarding me a bursary to travel to New York where I had the opportunity to meet Bob Power.

I must extend my gratitude to my countless students over the past ten years. Your inquisitive nature and desire to learn have deepened my own knowledge and allowed me to explore the topic from many angles and perspectives.

Lastly, I would like to thank my publishers and the editors I have worked with over the last three years at Taylor & Francis. Their belief in me by offering a contract and their ongoing support and flexibility allowed me to navigate this new path.

Introduction to the book

The quest for mixing mastery is a fascinating journey that opens us to a deeper awareness of how music and sound work. It alerts us to the wonder of sonic perception, and as we develop our skill, we harness an almost magical quality in being able to shape listeners' experience of time and space. This ability to engage with perception means that we can move people emotionally, encourage people to get up and dance to the rhythms, or to release emotion and energy in synergy with the sounds. As mix engineer, we have an important role in unlocking and enhancing the potential within the sounds that we are asked to work on. But many developing mix engineers fail to unlock and enhance this potential and often can diminish or destroy this energy through poor technique and mis-assessment of what they should be doing. Therefore, to be a successful mix engineer who can unlock this potential, we must possess a range of skills to ensure that we make the right assessment and chose the correct course of action. We must also have the skills and the knowledge of our tools to ensure that we enact that course of action correctly. Sometimes we need to have restraint and other time we need to act boldly; importantly, we need to have the ability to know when either is appropriate. Mixing is an art, and thus, it is quite impossible to learn it entirely from a book. The artistic part of mixing is yours to uncover through the process of developing your craft. This book offers you insights into the craft and through learning the craft it hopes to develop your confidence and intuition and to establish experiences when you are able to be in flow with the sounds and respond in these moments intuitively. We get to these flow states by developing our skill and establishing a workflow. The ultimate goal is to do most of your mixing in these flow states.

This book's audience is not beginners; it is for people that have started mixing and have learnt the basics. Now they are doing mixes but are not getting the results they want and have many unanswered questions. This book hopes to go a long way to answering many of these questions. When the answer starts to be 'well it depends' rather than stick with this cop out, or to simplify and force a simple answer, this book encourages developing analytic skill to find the answers within the music; for many times, the answer lies in the music itself. But there is skill in identifying that through listening and analysis. Often questions such as 'How do I know when the mix is finished?', 'How do I know whether I should be gentle or heavy handed?' and 'How do I know how loud I should mix?' can be answered by listening and correct assessment of the music. Listening (and critical listening in particular) is the most important element of mixing. The critical element of this means that we make assessments and decide potential courses of action. When we are approaching our listening in this way, we know what needs to be done and we can then go ahead and do it. This book is very much concerned with asking the question of why you are doing what you are doing. It is quite easy to learn about the basic concepts of compression and equalisation and turn knobs in our plugins. We can easily follow someone else's step-by-step approach, but unless we are carrying out these actions in connection with our analysis of the sound we are working with and an assessment of an appropriate approach; the results of our mixes are likely to be underwhelming. Every action in mixing should be tied with a need that has been identified through listening.

Part One is important in that it establishes the mental approach that this book will maintain throughout and the mental approach that I believe is necessary to complete successful mixes. It offers a four-part definition of mixing that reveals the way through mixing as a cyclical process of listening analysis and decision-making. It then introduces the concept that is central to this book, through listening the path through the mix will emerge. Part Two is the mixing bootcamp. There are basic skills and core knowledge that is required for all mixers. Before we can get into the more complex abstract critical and analytic place where a path emerges through listening we need to develop certain skills. It is intended to be worked through at least once in a linear fashion. The idea is that you take time with this, and rather than just reading, you are actually doing mixing practice as you go. This section comes with audio component that you can work through to test your skills. I often teach this as a 12-week course, so if you are looking for some structure you might take one chapter per week. Part Three is mixing time (finally!). In this section, after you have completed Part Two, you are invited to start doing actual mixes whilst reading this book. This section offers solutions and approaches to common issues or challenges in a mix. The ideas are that after you have completed your listening analysis of a mix you are working on, you will have identified issues and challenges. You can go to the chapter where your issue is discussed and apply one or more of the solutions offered. It is a problem-based approach that can be useful for many years as a reference point. However, it also continues your training in refining your problem-solving skills and increasing your tool kit of solutions to problems. After going through all three parts of this book, you should be feeling pretty confident in your mixing. Part Four considers information that is interesting and relevant as you continue to develop your skill in mixing. As with all crafts the learning never ends, tools and technology changes, new approaches are developed. This section includes relevant topics such as AI and mixing, spatial audio as well as a discussion on mixing whilst producing and many more interesting topics. This information is considered not necessary in order to be a good mixer but interesting and useful once you are establishing a mixing practice as further areas for contemplation and experimentation.

Confidence is imperative whilst mixing, and in many ways, it is the last piece of the puzzle. Once you have gone through this book and developed your skill in both the technical mastery of your tools, and developed your critical listening and problem-solving abilities when listening, the final element needed is confidence. By following the guidelines of this book, engaging with the concept of '*only doing processing when you know exactly why you are doing it*', and through practicing your ear training – you will be able to measure by comparative listening if the chosen processing approach has remedied the sonic issue. Then, you can confidently move through your listening and mix plan identifying issues and using comparative listening and targeted processing to check of the items on your mix plan. The moment you start second guessing yourself is the moment your mix starts to fall to pieces. This book aims to develop your confidence – in your skills, your knowledge of your tools, your ability to listening critically and correctly and your problem-solving skills. In developing your skill, not just in the doing but more-so in the analysis and assessment, you will develop your own workflow that will not be a step-by-step static checklist but a dynamic process of engagement with sound aesthetics, with technical application always connected to creative ends. This clarity of purpose and confidence in skill will ultimately deliver you to moments when you are mixing in flow. Which is the ultimate place of creativity where we mix intuitively and with confidence.

About the author

Photographer: Sven Serkis

Jane Arnison is an engineer, producer, composer and sound artist with a career spanning popular and experimental music. Her expertise combines a conservatoire-based Music Bachelor's degree, a degree in philosophy and a Master's in sound art. Jane's creative work reflects her diverse background, ranging from collaborations with major labels like Sony Music to mentoring emerging talent. Equally at home in recording studios and fine art spaces, she has developed large-scale immersive installations as well as commercial music projects.

As a lecturer in music and sound, Jane teaches at New York University Berlin and at FHNW Basel's Jazz Campus. She frequently leads workshops and speaks at conferences and festivals on sound. Jane is committed to addressing industry imbalances, advocating for increased diversity and career opportunities in studio production for aspiring producers from underrepresented backgrounds.

Part One

Get it right from the start

Summary

To define an approach, we first must establish a foundation, a definition of mixing. My definition of mixing has four components that will form the basis of all the information that follows in the book. This section takes a moment to articulate the elements of mixing. Often people don't establish clear goals when doing a mix, they are just experimenting and trying things out. No wonder then that a common question that new mixers as is how do I know when the mix is finished? In this section, through defining the elements of a mix, we set out first to establish the primacy of critical listening and to encourage decision-making. These two factors illuminate issues and creates goals which results in actionable approach that leads to finished mixes. There is also a discussion on the difference between mixing and production and ultimately a reminder that we are part of a chain of creative processes, and we bear responsibility to respect our role and serve the song first and foremost.

Chapter 1

What is mixing?

"....it is the art of balancing all the elements together"
"...the moment of refining and optimising"

The above quotes are common responses when I pose this question to new mixing students. It is great if you were also thinking along the same lines because mixing, certainly is these things. Whilst mixing can be the above things, it can and should be further defined. Let us consider what we are doing when mixing? (You might like to make a list of the things that you do.)

Mixing is not production

Before I go further we need to address something important. Mixing can sometimes be very hard to separate from production. I understand why this is the case – both production and mixing can involve utilising similar tools, audio processing such as compression, reverb, equalisation, etc. Whilst these two aspects of working with sound are certainly sometimes blurred, it is nonetheless helpful to try to delineate them. If you can't define the difference between them, you cannot be sure that you are paying attention to the correct details or making the appropriate decisions.

What is the difference between production and mixing? For me, the best way to clarify the difference between mixing and production relates to *sonic aesthetic*. Throughout the production stage we are still defining the aesthetic of the piece. Mixing might offer adjustments and refinements to that aesthetic, but it rarely is the place where large decisions are made about the musical or sonic aesthetic.

production = crafting sonic identity, finalising arrangement
mixing = optimising what is already there.

NOTE: I'm not saying that mix engineers don't also ask questions that producers do, nor that producers never engage in processes that are mixing, but the ability to know the difference between them is important. You might be mixing and producing your own work and so if you stay in the producer mindset you can miss important aspects from a mixing perspective. On another note, if a mix engineer makes significant aesthetic adjustments to the production they might get into trouble with the producer, or perhaps if the artist/producer wants a mix engineer to make significant adjustments that results in shaping the aesthetic then the mixer becomes a producer and should be having a conversation about production credits on the work.

What is mixing? 3

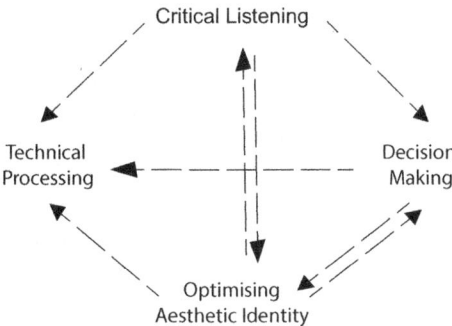

Figure 1.1 Mixing components and relationships

A working definition for mixing

In other words:

> Mixing is a repetitive process of critical listening & technical application, which, led by an awareness of the aesthetic intention, illuminates decisions about the sonic factors that need attention, what tools would be suitable and how to approach their application for optimal results.

Continue reading to the next section as we further analyse and expand on the elements of *critical listening, decision making, technical processing, optimising aesthetic identity* as defined above (Figure 1.1).

Critical listening with a mixing mindset

First up let's **pay attention to critical listening** because it is the most important facet of mixing. When we perform critical listening in a mix what are we doing? When we are mixing, we engage in a range of listening modalities. We often engage in a *discriminatory* type of listening where we must focus on specific elements and exclude other parts. For example, when we are trying to identify a ringing resonance in a snare drum recording. Also, when mixing we engage in *comparative* and *analytic* listening to reference tracks as well as *diagnostic* listening to the pre-mix. We do not need to articulate all the modalities (but you can in your own time if you want!), rather this should alert you to the fact that our critical interaction with the act of listening in mixing is constantly changing based on the needs of the specific task at hand. The ability to move between these modalities requires training and focus. The type of training necessary will be discussed in Part Two of this book.

We want to listen critically with a focus on details relevant to mixing. So, what are those details? In addition to developing the required skill to move between different listening modes whilst analysing sounds, there is also another layer which relates to the decisions that we will be making during our work. As suggested earlier the key decisions that a producer and mix engineer make are different. If we do not have our brains engaged correctly, we will not be listening correctly. We can focus our attention by posing leading questions to ensure we pay attention to the things that matter to us as mix engineers. Below are some example questions that illustrate this point by comparing the types of questions producers ask vs that of mixers. Being clear about the types of questions you should be asking helps to ensure that you are listening with a mixing mindset!

Questions from production perspective	Questions from mixing perspective
Does the snare sample fit with the 1960s sound I'm going for?	Does the snare have any resonant rings, how is the dynamic stability?
Does the vocal arrangement develop well through the piece, are the voicing of the harmonic progressions working well?	Does the vocal arrangement have problems with sibilance, is the stereo image of the vocal harmonies optimised?

Examples of questions I ask when listening to ensure I am listening with a mixing mindset

- What are the heroes of the piece?
- What is the aesthetic identity? Do I have references, and do they make sense?
- What is the energy of the track? Does it match the intention of the production?
- Are there parts with particular issues? Or is there any element letting the overall piece down?
- How is the low-end balance?
- Are there parts that are fighting for attention?
- How does the piece progress through time? Does it captivate our attention? How are the transitions leading us to the next section?
- Is the sound dynamic or largely restricted/compressed with small dynamic range?
- Can we class the instruments and the areas of the frequency spectrum they are 'primarily' occupying. Furthermore, can we classify describe the characteristics of the overall sound and consider the approximate frequency? For example, the bass might sit in the low frequency, but is there any sub-information? What about the mid-range? Is their knock or a honk resonance? What about the upper frequency is there any noisy/scratchiness of the pick hitting the string?
- Is there an instrument/group that seems to be in front in the mix – hint in pop music it will be the vocals!
- How is the stereo image managed? What sounds are mono? Which are stereo? What are panned? Is there phase/time-based processing to enhance or create a sense of stereo?
- What about the use of spaces in the mix? Can you identify reverbs? Can you tell whether they are plates or springs, chambers or halls? Can you tell if they are short long? What about delays?
- Are the transients pronounced or rounded? Is there an edginess when listening? what is the transient of the kick drum relative to the snare? And the vocals?
- What about space? Can you identify different spaces? On which instruments? Are there delays?
- Are there any dynamic/side chain process? Ducking, gating etc.
- What is happening in the low end? Is there one instrument dominating or a couple? And how is that managed? Prioritised?
- What is the primary 'energy' off the song – i.e. sing melody, move to the groove – how is this translated?

This is not a definitive or complete list – but it gives you a good sense of the questions that I often ask myself when I am analysing a production that I am mixing as well as a reference track that I am deconstructing. Some of these questions might not yet make sense to you. That is ok, you can come back to this after you have completed Part Two of the book where you engage in deep training! For now, the most important takeaway is this: **to help us listen correctly we can ask ourselves questions to focus our attention on the right elements.**

Decision making – from listening a plan emerges

Once we have listened to the piece we will be mixing, along with any reference tracks, and have answered specific questions, we gain a clearer understanding of what we need to do in the mix. The abstract concepts of mixing begin to translate into actionable steps. This is when we can start to develop a plan for our mix.

It's important to note that not every mix will require the same approach, so we must build our strategy based on a thorough analysis of the sound beforehand.

In the previous section, I provided an extended (but not exhaustive) list of questions that I ask myself when analysing a piece before mixing it. The answers to these questions form my mix plan, or, in other words, the goals I aim to achieve during the mixing process.

Example of a plan for a mix after critical analysis of the pre-mix of a pop song – problems identified and potential solutions suggested

- *No head room – gain stage approx –3 dB*
- *Cohesion needed – mix bus processing to glue elements together*
- *Issues in certain sounds – clear up problem resonances in drums (snare, toms) – surgical EQ and gating?*
- *Low-end muddy and difficulty to hear bass melody at all times. Get the kick and bass working better together – eq, comp, etc.*
- *The mix is a bit dull or plain – make the keyboards sound lush – wider and some bigger space and movement (del)*
- *Vocal lyrics are sometimes unclear getting lost in the mix – clip gain adjustments, vocal signal chain, parallel comp?*
- *The song lacks increase in energy – create a sense of 'through motion' particularly at transitions so that there is more excitement as the song builds*

Targeted technical processing

The plan outlined above indicates the types of processing I will need to apply during the mix. It is important to note that this plan is specific, not general. You might be wondering if there are similarities between different mixes and if there is a standard plan I can follow. Whilst you are indeed creating a plan through your listening, it is also true that most mixes share common overall processes that we can observe. It might look something like this:

- *Listening analysis*
- *Organisation of the mix*

- *Levelling and balancing*
- *Clean and fix sounds*
- *Consider the dynamics and maximise the power of the sounds, individually and collectively*
- *Work with spatial elements to create depth and width*
- *Consider the play through and the journey through the piece from start to end considering transitions and the rise and fall of energy*

Gaining a basic overview of the actions performed by a mix engineer is an excellent starting point. You can find this information in introductory texts on mixing. If you are not familiar with the general processes involved in mixing, I recommend starting with a beginner's book to practice mixing. Once you have mastered the basics, you can return to this text. The issue with simply following a generic list of mixing steps is that you may overlook the specific needs of the audio in front of you. By concentrating on general tasks instead of the unique details relevant to each mix, you may end up with a final mix that falls short of its full potential.

Optimising the aesthetic identity

We have reached the final concept: aesthetic identity. It is important for us to accurately capture the intended aesthetic of the producer and artists for this piece. This is likely connected to the identity of the artist as well as their overall intention for the sonic work.

What do we mean by aesthetic identity?

One way to understand aesthetics is to think of it in terms of taste or style. When we add identity to aesthetics, it takes shape as the specific character of music, including its style, sounds, instrumentation, lyrics and the delivery of both instrumental and vocal performances by the musicians. All these elements communicate a distinct identity or character, which is formed by the various sounds we hear. The style and character of a song can be understood as its aesthetic identity. If we misinterpret the aesthetic intention of a song, the artist and producer will likely be quick to express their dissatisfaction with our mix. Therefore, it is essential to feel confident in your understanding of the song's aesthetic intention and to consider this when deciding which tools and techniques to use in the mixing process. For example, the mixing approach we take for techno music is vastly different from that used for jazz. These two styles require entirely different techniques regarding loudness, dynamic control and the spatial separation between tracks. If we apply a techno mixing style to a jazz piece – or vice versa – we would likely end up with a poor mix. When considering the aesthetic aspects of a mix, potential approaches may be more nuanced.

Revision of this section and moving on to Part Two

It is essential to understand what mixing is and to distinguish it from other similar tasks. Mixing involves not only listening but also engaging with a specific aesthetic, making decisions and applying technical processing. Critical listening is particularly important, as it helps us identify the specific processes our mix will require. Through careful listening, we can assess whether our processing enhances or detracts from the mix.

Recognising what is needed for a successful mix is the first step we have completed. Now, we will move on to the crucial training! The previous statement assumes a high level of skill. To effectively conduct a listening analysis, we must train our ears. If we do not develop our knowledge and skills effectively from the beginning, our listening may be inadequate, leading to a flawed mixing plan.

To execute the necessary processing and select the appropriate techniques for specific sonic challenges, we need a deep understanding of our tools and ample practice using them.

Part Two

Mixing bootcamp – getting the most out of our tools and ourselves

Summary

The process of mixing is not as mysterious as it may seem. While it's relatively easy to learn about using compressors and setting up for a mix, the challenge lies in creating a mix that sounds exactly how we envision it. Part Two of this book focuses on the necessary training to develop the skills of a top-class mix engineer. I have organised this section chronologically, reflecting the typical progression many people experience through these topics.

First, we will explore the tools available to us. In an era where there is an abundance of excellent tools, it is essential to know how to approach them effectively to enhance our mixing rather than complicate it. We will consider the theoretical aspects to encourage critical thinking, followed by practical guidelines for experimenting with our tools. You are encouraged to take this section in a modular fashion - one group of tools at a time. After the theory section then move onto the practice section, before returning to the theory and so on. The ultimate goal is to assemble a personalised toolkit of audio processing tools for mixing.

Next, we will examine our most valuable asset: our ears. We will discuss ways to train our ears to become critically sensitive to the details that a mix engineer needs to notice. Once we have developed our auditory skills and chosen our tools, it might seem that we are ready to mix. However, we must also consider the external factors that can impede our progress, such as the acoustics of our room and the quality of our monitors.

Finally, we will conclude with a discussion about the importance of following these strategies and guidelines, ultimately aiming to achieve flow states while mixing.

Chapter 2

Interacting with the multitude of tools

How do I know which tool to use when and why?

The question above is one of the most common inquiries from developing mix engineers. The honest answer may not always be what we want to hear: you need to practice. Over time, you will develop an intuitive sense of which tool is best for each job. Until you reach that point, you can choose tools based on expert advice whilst also allowing *yourself to* experiment and make decisions using deductive reasoning.

I've noticed a lack of guidance on how to learn about these tools in a way that fosters an intuitive ability to select different tools for various tasks. In the following chapter, we will embark on a detailed exploration of the key processes involved in mixing, such as compression, equalisation, saturation and more. Whilst discussing these concepts, we will focus on how to familiarise yourself with your tools and determine if your current options are sufficient or if you want to expand your mixing toolkit.

As a mix engineer, compressors, equalisers and other sound processing tools are akin to our paintbrushes. They are the means through which we create and craft a beautiful mix. Therefore, before we can achieve mastery like Michelangelo, we must do what all great artists have done: immerse ourselves in technical training to learn about our tools and practice using them. This section aims to help you approach these tools with curiosity and interest. Instead of randomly adjusting settings based on someone else's preferences, I hope to encourage you to ask yourself 'WHY', rather than just 'HOW'.

Whilst we can't cover all tools in detail – there are literally thousands of variations, with new plugins being developed daily – this can be overwhelming for mix engineers, both new and experienced. My approach is to categorise the different types of tools, considering their benefits and drawbacks in a comparative manner.

Ultimately, the goal of Part Two is for you to create a personalised mixing toolkit of audio processing tools that suit your style and that you have taken the time to learn in depth. This will help you understand why you might choose one tool over another in a given mixing context.

Now, let's get started!

Equalisation

When people think of mix engineers, they often picture someone operating a large, complex mixing console. Although this might not reflect the current reality of mixing, the

Figure 2.1 Analog mixing console

association with the console is still strong. So, what are all those knobs used for? As it turns out, much of the console's space is occupied by filters and equalisers. This emphasises the importance of equalisation – frequency adjustment and balancing – during the mixing process.

Mixing consoles are quite an expensive investment, so studio owners typically take time to make a decision before purchasing one (Figure 2.1). The equalisers built into these consoles serve as the primary equalisation tool for mix engineers. There are several key manufacturers of mixing consoles, ranging from well-known brands like SSL and Neve to more niche companies like Camarq and Harrison. Each of these consoles offers a unique sound character or 'flavour'.

The flavour of console equalisers is influenced by a few key concepts that are important to understand. First, because consoles are made from physical materials, the various components affect the overall sound. There are several elements to consider: the type of circuit used in the equipment, the types of transformers employed and the controls or settings that can be adjusted on the different EQ bands. After deep diving on these aspects the conversation will move to a discussion of AI equalisation processing before finishing up this section with a moment to consider phase and equalisation.

Different equaliser circuits

Equalisers can be broadly organised into two groups. Passive and Active.

Passive equalisation

What are they? The filter circuits are simple and have less components, they do their job without transistors, op amps and rather use passive components such as resisters, inductors. Now if you are not knowledgeable on in electronics it is fine you don't need to know all this detail. All you need to know is that it is using **passive components, and less parts are used to complete the circuit.**

What are some famous types?: Manley Massive Passive, Pultec EQP1A

How do they effect the sound? This is the important part! They are made up of passive components, i.e. no need to power up the circuit. There are less components in a passive circuit, and so generally, they do less 'harm' to the audio. They will introduce less artefacts, colouration, distortions. Also, the inductors will create musical adjustments that can create a richness, gentle saturation. Low frequencies can become thicker, and higher frequencies can be adjusted without introducing harshness. Conversely harsh high frequencies can be reduced without taking away the life from the signal.

Some examples of good uses? These are often used in mastering signal chains, which makes sense because of their clear and non-distorting transient quality. In a mix, they are great for mix buses and also subgroups. But you might also like to use them on parts of your track that need adjustment without any phase or transient diffusion, i.e. low-end parts such as kick and bass often love them as well. I love putting them on harsh vocals to calm down the top end without ruining the energy. And I love using the Pultec EQP1A with the cut boost technique on the low end to tighten and boost at the same time (look it up!)

Limitations? You will definitely not use a passive EQ to remove the ringing resonance of a snare drum. They can be too subtle for significant tone shaping.

Active equalisation

What are they? The term active equalisation is not really used commonly so you may not have heard this before. Most equalisers made after transistors were invented are technically active, such as all our famous console equalisers from SSL, Neve and API.

What are some famous types? API 550, SSL 611, Neve 1073.

How do they effect the sound? Active EQ's tend to have a sharper sound. This is often due to their adjustable Q controls which allows for more precise and exaggerated filters. The active, or powered nature means that active eq's have the ability to significantly boost the signal. Consider the SSL G series console eqs have up to 18 dB of gain boost or cut, we can rely on them for dramatic adjustments to the tone.

Some examples of good uses? These were the workhorse eq's of old school analog large format console mixing. So they did the bulk of the EQ work in the mix. They are great for removing ringing resonances with a sharp q. They are also great for creating space in the mix between different instruments. They also can dramatically change the overall tone of the sound. Each brand/maker of these console eq's has its own sound based on the design of the circuit. So it's a great idea to investigate in detail the special characteristics of the console eq's that you are drawn to use in your mix. For example, the famous high shelf of the Neve 1073, or the power of the low cut on the SSL.

Limitations? These can add more artefacts, subtle but noticeable, They can make massive changes to the tone very easily. For these reasons, they might not be the best solution on your mix bus.

What about digital equalisers?

The first thing to note, is that these days pretty much everything that originated as hardware is also available as software or digitally. We treat these software emulation versions as if they were analog hardware, because they have been created to emulate so accurately the behaviour of their physical inspirations. So for all of these equaliser emulations see the above two sections.

There are however equalisers that are digital in creation and concept and sound and work differently to both the active and the passive equalisers we discussed already.

What are they? Digital equalisers that use algorithms, or maths equations to carry out their processes. Rather than a physical circuit.

What are some examples? Fab Filter Pro Q 3, Waves Q10. The standard equaliser in your digital audio workstation (DAW) is most likely an algorithmic design. Such as the channel EQ in Logic, or the EQ Eight in Ableton.

How do they effect the sound? The characteristic of digital EQ's is that they are precise. They tend to have more flexibility with more controls. They are generally 'transparent' meaning that they add no additional colour to the sound through transformers, tube output gain stages or opamps in their circuit.

When are they used for? Digital EQ's are great for cleaning up sound, They are great for removing resonances, and as they often come with a spectrum visualiser, they are helpful for those who like visual aids (however proceed with caution here, always trust the ears over the eyes!). These days they have often replaced the console equalisers that were the workhorses in the analog days, in a digital context these are often the first equaliser that you will reach for.

Limitations? Digital eq's are clean and transparent, as it turns out this is not always what we want. Over-use of these types of EQ's can result in boring mixes. They don't aid in adding any 'character' to the sound that has been processed.

Transformers colouring sound

Transformers are a first indicator of the colour that a certain piece of gear may have. For example, the SSL channel strips (and console EQ's as part of them) have a different sound to the Neve channel strips. They all use different transformers. Furthermore, the same manufacturer creating the same piece of gear in different time periods might use different transformers. The 1970s version of the Neve 1073 sounds very different from the 2024 version of the same unit. This is at least in part due to the fact that different transformers have been used.

So what is a transformer anyway? A transformer most simply transforms an audio signal in some way (Figure 2.2). Most commonly they change the impedance, or the resistance of the circuit. You don't need to worry yourself too much about what it does unless you are interested, but what you should be aware of is that transformers are common components in much famous gear, and the transformer can play a huge role in the sonic imprint of that gear. You might have come across transformer plug ins (SSL, Kush) and thought what are they for? Well now you can add the colour of these components into your Digital signal flow

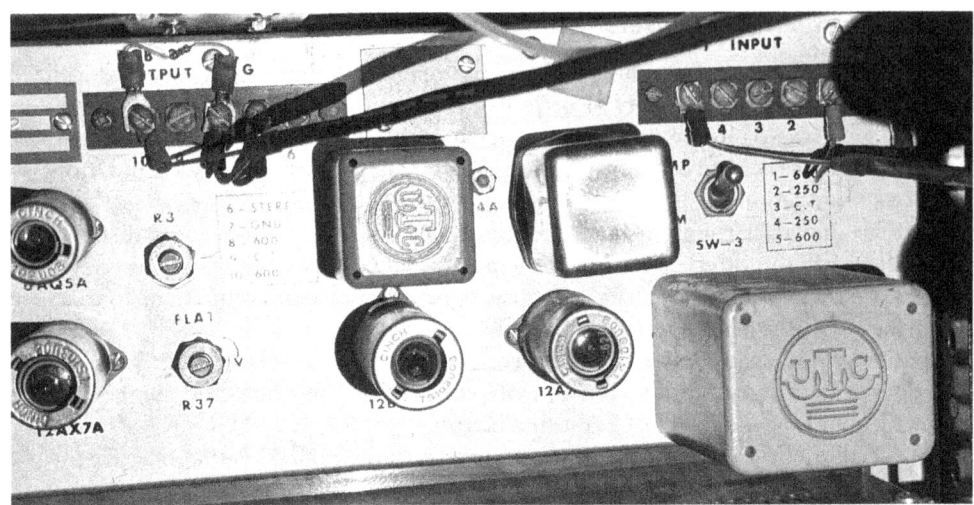

Figure 2.2 Transformers in the back of audio hardware

Controls

Another aspect to consider with EQ's and in particular with hardware modelled EQ's are the controls. If we compare three different console style eq's SSL 611, API 550 and the Neve 1073, you will notice that the exact makeup of the eq's is different. Some have hi-pass filter and others don't, and whether it is a shelf, pass or a notch/bell filter, they will often be slightly differently designed.

As a developing mix engineer, it is a good idea to familiarise yourself with some key differences. Below are some questions you might like to ask when comparing EQ types.

What slope is the lo/high pass?

For example, the SSL 611 EQ lo pass is a slope of 18 db/8va whereas the Neve 1073 has a slope of 20 db/8va (Figure 2.3).

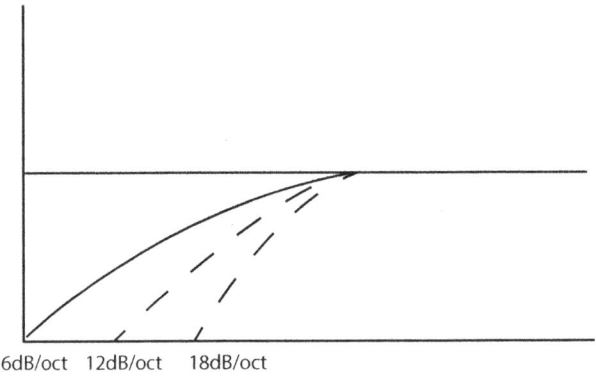

Figure 2.3 Different EQ slopes

Interacting with the multitude of tools   15

What is the maximum gain adjustment?

We can just compare two types of the same EQ for this example. The E series (brown) SSL has 15 db boost/cut, whereas the G series (Black)18 db – so this means that the G series EQ has more dramatic eq adjustments for the same physical movement of the knob.

What frequencies can be adjusted?

In the world of digital equalisers, we are used to being able to set it to anywhere within the 20–20 kHz range, but specific hardware models have fixed frequencies that can be set (Figures 2.4a and 2.4b).

Let's compare the Neve 1073 and the API 550 in this example.

The neve 1073 has low shelf at 35, 60 110 and 220, and a notch at 360, 700, 1.6k, 3.2k, 4.8k and 7.2 k.

The API 550 has a low shelf with settings at 50, 100, 200, 300 and 400 and notch filter at 400, 800, 1.5k, 3k and 5k.

Why are these settings different?

Neve say that the 1073 frequencies were selected for optimal use in a music recording console to be able to perform across a wider spectrum of sounds than if it were used

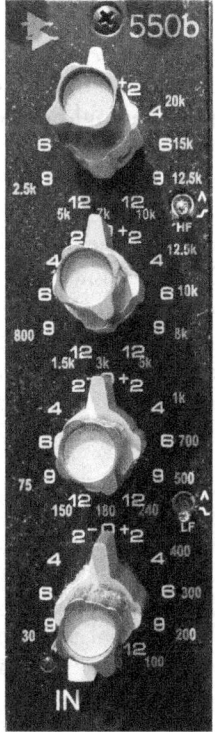 

Figure 2.4 API and 1073 Equalisers

for the broadcast industry where the focus is mainly on the speech spectrum. It is often deemed by the manufacturers to be areas where common build up and problems occur, or conversely where sound benefits from boosting. Ultimately, it also helps carve out the sonic identity of the different units and brands.

Why would knowing this impact your decision to use one or the other?

These are just a couple of questions to get you started, you might start to formulate your own as you start digging in more to the various hardware style eq's available for you to use when mixing.

- Knowing the power/strength of slopes and gain boost and cut is helpful for the style of the mix you are doing. If you want something that is punchy and edgy, the G series console eq with extra db boost/cut might give you more dramatic results
- If you need to carve detail in the low end, the Neve 1073 is not going to help you, you will likely consider the SSL or the API style console EQ's for this

AI and adaptive equalisation

What are they? Digital equalisers that utilise AI algorithms that can perform real-time analysis and make adaptive adjustments and automated responses to the incoming audio signal.

What are some examples? iZotope Neutron, Sonible.

What are they used for? These EQ's use adaptive algorithms or AI to analyse and interpret then adjust your audio signal overtime. They are great for controlling resonances, or unwanted artefacts. They are great for the kind of cleaning up and balancing. Using AI instead or impulse responses in analog emulations. Transferring eq shapes of one sound source onto another. As time progresses we will see more integrations of AI into our processing workflows.

Limitations? They are used in many different ways, so it is difficult to pinpoint limitations in a broad stroke. Adaptive EQs like Sonible or Neutron don't offer a colour or a style that analog style EQ's do, they are more akin to digital equalisers. The main issue however for working with AI tools is the tendency to stop critical attention when using automated processes is there, not that this is the plugin's fault but it is important to remember that the mixer must still exercise their discretion.

Phase and equalisation

When I first introduce the topic of phase and equalisation to students, there is understandably much confusion. We consider equalisation to be 'out of time' so to say unless we are working with complex combination processing such as dynamic eq. All audio processing takes time, and even though the impact on the signal is minimal, it can be enough to introduce phase alterations. The amount of phase alteration that an EQ will introduce is dependent on the circuit design, as well as the type of filter, and how extreme the settings are. Far from being a problem, the subtle phase adjustments, can create part of the EQ's style and character and we can use the phase artefacts creatively in the mix.

It is important for you as a mix engineer to get curious about phase in relation to working with EQ.

Below are some tips and thoughts for helping you when exploring this concept.

1. The more extreme the filter q or shelf slope the more intense the phase distortion
   - Be aware of this when you want to cut out low frequencies – for example if you use a sharp cut on a reverb eq you might pronounce some parts near where you are trying to lessen the presence of tone
   - Conversely when you are working with sub-frequencies, you might use a sharp cut to drastically reduce signal below a certain point but if you get the slope positioned in the right place, it can have the effect of feeling like you have increased the bass response. So, you have reduced the low energy and tightened up the low end, but given a boost – great!
2. For **surgical eq** when you are wanting to cut specific ringing resonance, it might be an idea to try a linear phase – this will provide more clear results, but be careful to listen to the introduced artefact of pre-ringing
3. As you get into top down processing – subgroups, mix bus, mastering chains – pay particular attention to what EQ you are using, and listen to whether you are weakening your signal through your approach to EQ – it might be introducing phase

------ Pause reading here and go to Chapter 3 – FROM THEORY TO PRACTICE before reading on. Here, you will put all the information discussed in the previous pages into action exploring and playing with various equalisation tools. Ultimately, you should select your tools and learn them ready to put them to use when needed in a mix. ------

Compression

Compression is one of the most discussed and often misunderstood tools for aspiring mix engineers. Whilst the concept of compression is straightforward, its application is subtle and complex. Mastering compression begins with understanding its various components and how we should think about sound when applying it. This involves learning about different types of compressors, how their circuits respond and how adjusting their settings impacts the sound. Armed with this knowledge, you can practice using your chosen tools and develop a more refined ear for the nuances of compression.

More than just level balancing

What is compression really? Beginners would answer this by saying dynamic range control. So basically, using a compression is about levelling or adjusting the amplitude of a signal. Whilst this is true, if you stop there with compression however, you are missing the full picture and may well be making errors.

The most important thing to understand is that you must always consider **time** as well. The dynamic fluctuations through time. Compression is a tool that impacts **amplitude over time**. Which can also be understood as amplitude envelop shaping. It isn't **just** turning down certain sounds or reducing the dynamic range of a sound by reducing the peaks by a certain ratio. We must also consider the way that it turns them down as these dynamic fluctuations are changing constantly over time and the compression circuit and settings will affect this in different ways effectively changing the envelope/shape of the sound.

If we take something repetitive like a kick drum, we might be able to hear how the compressor can drastically change the character of a sound buy adjusting its envelope. Depending on how we compress it we might bring out the click transient part of the sound, or we might accentuate the drum resonance or body of the sound, we could also accentuate or elongate the release or vice versa and tighten up the release to make the drum sound tighter.

Let's consider a more complex sound like a lead vocal. The way we set up the compressor as well as the type of circuit that we use can have big impact on how the audience interprets a layer of sound. For example, if we use a very fast compressor it will help accentuate the consonants and punctuate the rhythmical qualities, which is great for an energetic song, but not good for a ballad. Conversely if we use a very slow response compressor circuit/setting on a song that needs energy and movement it will destroy any sense of movement and blend everything together which will result in the part likely feeling sluggish.

The above examples are meant to articulate a few important points. Firstly, when considering compression, we will be thinking of more than just level balancing. Secondly, we need to have a good working knowledge of various types of compressors. Thirdly, we need to understand the various settings, that either are built into a compressor type or that we can adjust ourselves.

Finally, it's time to then put all this together and start building mastery with compression.

Various types of compressors

The big five circuits are OPTO, Tube/VariMu, VCA, FET and Digital (Figure 2.5). It can help to consider the type of compressor as it also often shares characteristics that can help you to make decisions on which compressor is appropriate to use for a given processing task. Let's go through the circuits in a roughly time-based list starting from oldest to youngest.

Figure 2.5 Various types of compressors

Opto compressors

Example: LA2A

Description: Opto compressors utilise a photo-resistor for the reduction part of the circuit (Figure 2.6). As input amplitude increases an LED in the circuit shines more brightly which is picked up by the photo-sensitive resistor. They are known to be slow and smooth. Most often there are not many controls on opto compressors with the threshold being fixed and the attack and release dependent on the components used.

Opto compressors are often non-linear, and the famous LA2A has a two staged release phase (learn more by reading the manual!)

Characteristics: Warm, slow blending, non-linear

Popular uses: Ballad vocals, Creating cohesion to bass lines, strings and harmonic elements needing some focus without pushing forward too much.

Tube compressors

Example: Tube-Tech

Description: Tube compressors contain, you guessed it, tubes. So that immediately tells us something about what we might expect. Tubes are a bit slower in their response times, they can create some pleasant musical thickness and will round out spiky edges.

Figure 2.6 Optical circuit

Characteristics: Smooth. Responsive without being spiky Create some richness and thickness. Often there are not many variable controls, attack and release control only sometimes present. Often the tube part of the signal is responsible for the makeup gain, and the gain reduction is often opto.

Popular uses: Can be good for certain types of music, jazz, classical. For mix buses on more relaxed music. Great for lead vocal chains.

VCA compressors

Example: SSL G Bus

Description: Also famously known as the 'glue' compressor, VCA compressors officially stands for Voltage Controlled Amplifiers, which relates to the way the circuit works in a feedback configuration.

Characteristics: More transparent than tube and opto, responsive without being too edgy, create a sense of glue or fusion of parts.

Popular Uses: Mix bus! Subgroups. This compressor isn't called the glue for nothing. It is a great tool for creating a sense of cohesion between parts that need to work together, for example individual drum elements that want to be heard as one layer.

FET compressors

Example: 1176

Description: Field Effect Transistor are characterised by their spiky character hard knee and fast responsiveness. They have very fast attack and release, which are normally also present as controls that the user can adjust.

Characteristics: These are characteristically edgy, in that they generally have a hard knee, and they are not smooth. Fast and responsive.

Popular uses: Edgy sounds with attitude, great for fast moving music, for creating and enhancing movement. Great for fast vocals, e.g. rap. Great for percussion sounds and sounds with transient information. Can reintroduce movement to sluggish sounds with correct settings.

Digital compressors

Example: Fab Filter, Ableton Compressor, Logic Platinum Compressor

Description: Digital compressors use algorithms rather than circuits. These compression algorithms can model the behaviour of other models very well.

Characteristics: Transparent control without adding artefacts or colour. Can be very fast or very slow depending on the setting, Extreme gain reduction capabilities. Very flexible and adaptable due to the many controls that can be manually adjusted by the user. Often can dial in sounds that can mimic the circuit characteristics of the above styles that we have covered.

Popular uses: Go to standard compressor for many in the mix. Don't necessarily add too much character so good for a general control that doesn't want any extra additional colour.

One compressor to rule them all?

Is it possible that a digital compressor could offer everything that you might need and want. Yes, it is possible, to a certain extent. The logic compressor with its variable circuit types has been around for years and years, and only since the new graphic interface do many see them as discrete compressors. The Dutch company Fab Filter, have one of the most popular one stop compressors on the market. With their multitude of settings, many say this is the only compressor you will ever need.

For me though, I still like to have a tool kit of different compressors. There are a couple of reasons for this. Firstly, mixing is for me a creative experience, and part of my aesthetic links with analog approach and so I like to work with these tools. However a larger reason is probably the immediacy of getting 'the sound' that I'm going for. When I chose the Tube-Tech or the 1176, I immediately start with something approximating what I want. I make a couple of adjustments to the available controls to dial in the sound. With a digital compressor there is a lot more fiddling with the parameters, which is likely to take me out of the mix. This is subjective and of course each to their own, but at least some food for thought in considering what approach will suit your style best. Are you a one compressor kind of person? Or like to build a tool kit like me?

Uncovering the secrets of our tools

This book assumes a basic knowledge of how to use a compressor and an understanding of the main knobs and their functions. However, if you have only worked with digital or DAW-based compressors, you might not have explored some of the more complex variables found in analog compressors. Many of these variables are fixed, meaning they cannot be adjusted. Understanding these settings helps us grasp the characteristics of specific compressors. For instance, the FET compressor has a hard knee that is not adjustable, whilst the LA2A features a soft knee. The 1176 has a fixed threshold, whereas the VCA allows us to lower the threshold of the circuit.

Now, beyond reviewing the basic elements of compressors, let's delve into these concepts a bit further and uncover some of the more specialised and advanced topics. I will provide some insights and tips, but I do not intend to deliver an exhaustive analysis of every detail of every compressor ever made – this is obviously impossible! The aim is for my notes to serve as a springboard for your further research, encouraging you to read manuals and experiment with the actual processors. By doing so, you can build your mastery of each process tailored to your own tastes and tools.

Attack and release and knee

We should know that *attack* is how quickly the circuit activates after a signal has reached the threshold.

What you might not know: *Knee* is less obvious. The knee tells us how quickly the circuit arrives at the maximum ratio after the attack time has passed. So a hard knee goes directly to the maximum ratio (say for example 4:1), a soft knee will take a bit more time to scale up to 4:1. Ok, so what does this mean in sonic terms? It has an impact on the general style or aesthetic of the compressor circuit – a hard kneed will be likely to be rougher, more abrasive and perhaps more artefacts likely. A smooth or slow knee will be smooth and more gentle. Without reading manuals, often for analog gear you do not know what the knee is. But you can probably guess if we compare the fast and abrasive 1176 compressor to the smooth and creamy LA2A?

How do you set your release? I like to set my release in relation to the needle on the meter (which is often a VU meter in an analog compressor). It is a bit weird but I'll share it with you now! I imagine the needle is someone dancing to the music. So techno the needle goes up and down very uniform, for house and then funk the needle has a bit of a curve or slower response…for ballad it is slower again.

Now whether you like my analogy or you have your own way of working with it, it is something that most people get a handle on (at least conceptually) from the beginning. You might not know more specialised details such as LA2A and all optical compressors for that matter, have a two staged release character. I don't want to spoil the fun for you – get into the research yourself and discover more about how this works, and how you can use this characteristic to great effect in your mixing.

Threshold

What you should know: This allows us to set the point at which the compressor should start working. So based on the loudness of our signal we lower the threshold until we start seeing gain reduction (more on that later).

What you might not know: What about a compressor like the 1176 which has no threshold? For compressors like this one, the threshold is FIXED and to effect gain reduction we must increase the level of the audio signal coming into the compressor. Which means that we need to pay close attention to the makeup gain to ensure we maintain the balance of the compressor on the way out.

Gain reduction and threshold

I've mentored many people on their mixing techniques, and I've often noticed that they aren't effectively using their compressor settings. Whilst it's perfectly fine to use presets,

I want to stress that you must always adjust the threshold and determine the amount of gain reduction you want, even when using a preset.

You can monitor this using a level meter, such as a VU meter in analog gear. Understanding gain reduction is crucial when compressing. Even beginners using presets need to manually set the threshold to achieve their desired level of gain reduction.

It's important to become familiar with identifying the amount of gain reduction happening then adjust the threshold (or input gain) to reach your target amount. Additionally, keep in mind that the attack and release settings will also influence gain reduction based on how they respond to different parts of the signal.

What is considered a little or a lot of compression? When I ask my students what is considered a lot of gain reduction, the answers vary greatly!!

Gentle Compression: 1 or 2 dB
Medium Compression: 3 to 6 dB
Strong compression: 6 dB and above

Makeup gain

People often don't think too much about makeup gain. It just turns up the level post the reduction introduced by the compression. But it turns out, when we are considering analog compressors there is a more to the story. With tube compressors in particular, the makeup gain is the place where colouration and character are introduced. In fact, if you wanted you could turn your simple tube compressor into a tube saturation unit by driving the output!

The magic of manuals

I've just shared a few little secrets that some of you may not know, but there are many more! It's too much to write down here, but they are documented and easily accessible. The manuals for our favourite analog gear are available online. You'll find a wealth of exciting information in them. For example, the details of the two-stage optical release function on the LA2A can be explored, as well as how it works similarly or differently from the opto-tube compressor Tube-Tech CL1B. Another valuable resource is the Empirical Labs Distressor manual, where you can discover insights about the relationship between the ratio and the variable knee settings, along with many other great details.

> The fastest way to get mastery of your gear is to know what all the buttons do, understand the basics of the circuit design, and how the physical (or emulated) parts interact with and impact the sound. Then, you can experiment with this knowledge and connect the dots a great deal faster than random knob twiddling alone.

Other dynamics processing

Though they are one of the most utilised tools in mixing, compressors are not the only type of dynamic processing available. People often think of dynamics processing as just the vertical or amplitude axis. However as already discussed, the often-neglected element of dynamics processing is the envelope shaping. All loudness is experienced as an envelope, with an onset and an offset. Dynamics processors have a significant role in crafting the contour of individual sounds as they sound out in time. When we consider dynamics from this angle, it widens our frame when considering dynamics processing.

Gates and expanders

These are very useful tools. For example, with recorded sounds, that might need to separate the direct from the ambient sound recorded (example drum recordings). Gates also have a great use for shaping samples and for creating relationships between sounds in genres such as techno/house. Rather than choosing the Eq or compressor immediately to help a kick drum 'tighten up' you might try applying a gate and be surprised at the results!

Do you understand the difference between a gate and an expander (downward)? A gate is an infinite expander expanding the dynamic range downwards to 0.

Dynamic equalisation vs multiband compression

These two processes are closely related. They both combine selecting a frequency range as well as dynamic adjustments. There are some key differences that being aware of will help you decide which tool might be best for you.

Dynamic equalisation

Works with equalisation filters – notch, shelfs, etc. and applies a compressor to dynamically reduce the gain of the specific filter, maintaining the shape of the filter, be it a sharp q notch, or a 12 db/oct shelf. This can be great when there is a specific frequency in an instrument that rings out sometimes. For example, bass guitars, might have been performed with different finger pressure and occasionally one note rings out, but if you use a static eq it takes away that note all the time. In this case, the dynamic eq is perfect for the job. You might also use a dynamic EQ to reduce harsh frequencies, like the scratch of a finger on an acoustic guitar string. It will only apply the filter when this troublesome sound occurs and pushes past the threshold point at that frequency.

Multiband compression

Comes at it from a different angle in that it splits the frequency range into bands. Often you can determine how many bands and the size of the bands yourself. A standard approach in mastering was to split the frequency into four bands, but these days it is more nuanced and flexible than that. If we have set up four bands, there will then be a compressor on each band. For example, sub, low, mid, high. The compressor will listen to sound within that band (e.g., low = 85–250 Hz) and apply the compressor to the band equally.

When to use them?

Dynamic EQ is more likely to work with specific frequencies within one sound. It is great for targeting problems that are not always persistent so that when the sound is not there no processing is applied. And the reduction only comes when needed.

Multiband compression is a broader approach and thus is more commonly used for adjusting groups, so a drum bus, or a vocal bus or even the mix bus might have a multiband compressor.

With both tools, as with all tools, only use it if you really know **why** you are using it, and if the standard approach of EQ and compression is not working. Depending on where you apply the processing, both tools can have quite a dramatic effect on the sound. Use them with caution!

Sidechaining

We complexify the situation when we start to consider sidechaining. This process takes the audio signal from another part in the mix and uses that to activate a dynamics processor. This was made famous by artists such as Daft Punk with their technique of taking a kick signal and using that signal to turn on and off a compressor on a synth or bass part. This technique known as ducking is now a common element in much dance music.

This process can be used more subtly in mixing to help make space. I will discuss this more in detail in Part Three of the book, but for now it is important to take time to explore the concept of sidechaining as you gain mastery of your tools and find out what of your dynamic tools allow you to 'key in' a side chain signal?

Side chaining is it worth the hype?

Whilst side chaining is a very useful tool, I believe it is often misused and can lead to lazy mixing and lacklustre results. I've noticed an increasing trend of producers resorting to these techniques without taking the time to listen and accurately diagnose issues in their mixes. Instead, they side chain everything, which can create several problems. One issue is that side chaining can create an unintended bonding between two elements, sometimes having the opposite effect of what is intended. This reliance on side chaining often means that critical listening is neglected, which can lead to more issues down the line. Another significant reason I disapprove of automatically reaching for side chaining to 'clean and clear' the relationships between parts is that it serves another important purpose: enhancing movement and groove. In a forthcoming section, I'll discuss this in more detail. If side chain compression is used for EQ balancing, it becomes less effective when trying to create movement. More commonly, this aspect of side chaining is overlooked entirely.

------ Pause reading here and go to Chapter 3 – FROM THEORY TO PRACTICE before reading on. Here, you will put all the information discussed in the previous pages into action exploring and playing with various equalisation tools. Ultimately, you should select your tools and learn them ready to put them to use when needed in a mix. ------

26 Mixing in Flow

Saturation and distortion

Saturation is a much talked about technique used in mixing but grasping exactly what it is and how it is different from other types of signal distortion and 'drive' processing can be difficult. It gets even more confusing when we consider the different types of saturation such as tube and tape. In this section, we start by defining common terms that pop up when we engage in discussions about saturation and distortion.

Some definitions

Waveshaping: Though we are using different names for these processes which describe their sonic character, strictly speaking all dynamics processing as well as the following section can be understood as WAVESHAPING. Wave shaping most simply describes transforming an input waveform (i.e. input signal) by applying a non-linear function (i.e. an exponential sine function).

Distortion: This is a very misleading term. It is one of those catch all terms that can mean anything! Therefore, it often tells us less than we want to know. Distortion means change on the base level and generally has an association that change occurs in an unintended or undesirable way. We can have phase distortions occurring when EQing. We can also have harmonic distortion through waveshaping. We can also have other forms of digital distortions such as aliasing or quantisation.

Non-linear function: simply means that the output signal is not proportionate to the input signal. A linear function applies a change equally to all parts of the signal, for example raising the volume is a linear function. Non-linear means that part of the signal is changed, so this might be through clipping for example where the loudest parts of the signal are pushed down or clipped through the processing.

Hard clipping: Anything louder than the threshold is abruptly and harshly truncated or clipped. This creates a horizontal line at the threshold point as it reduces the signal amplitude (Figure 2.7). This significantly changes the signal's timbre introducing upper frequencies or harmonics.

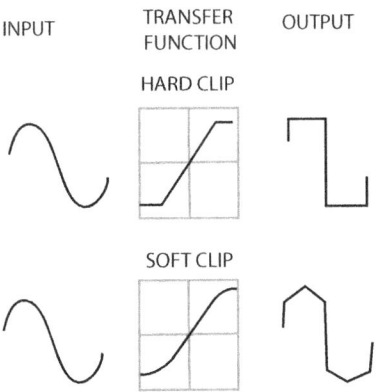

Figure 2.7 Clipping wave shaping diagrams

Digital clipping – anything louder than 0 dB FS peak is automatically truncated or stopped at 0 dB, this gives us something resembling a square wave, this abrupt cutting off the flow of the sound wave, creates a sharp edge, like a square wave, pronouncing the edge of the transient. The sound is harsh and noisy with both even and odd harmonics present. This is very similar to hard clipping, it could be viewed as a type of hard clipping. It is different from hard clipping in that it generally occurs as a side effect to poor gain staging, it is only available in the digital domain and it only relates to the maximum threshold being breached. Therefore, it is harder to control.

Soft clipping: Is a form of non-linear waveshaping that occurs when a signal is driven in certain systems past the threshold point. The limiting of the amplitude occurs in a gradual or more gentle manner, more like a type of compression. It rounds off the peaks rather than harsh cut that occurs with hard clipping. It results in a smoother more musical type of distortion.

Harmonic distortion: occurs when non-linear changes occur to the signal. These changes create new frequencies that are integer multiples of the original signal. These are called harmonics.

2nd order: 2, 4, 6 overtones – these are octaves and 5ths and therefore create a musical enhancement to the signal without introducing harmonic dissonances. Often referred to as musical or warm. If this doesn't make sense, now is a good time to research the natural harmonic series and find out about the overtones.

3rd order: 3, 5, 7 overtones – more complex and therefore more blending

Saturation – occurs when the audio level exceeds the maximum capacity of the system. Systems that produce saturation artefacts include tape machine, tube amplifiers and transformers found in analog gear. It creates harmonic distortion. Depending on the system makeup it may favour 2nd- or 3rd-order harmonic. The resulting distortion creates soft clipping.

Tape saturation – generally favours even-order harmonics. Creates a compressed and glued feeling to the signal. Can introduce more obvious destruction of the signal integrity depending on the tape and the speed. Introduces artefacts such as tape hiss, as well as wow and flutter due to the mechanics of the tape.

Tube saturation – generally favours even-order harmonics. Creates a sense of cohesion through the compression characteristics. The tubes can also produce odd harmonics when driven. The types of unwanted artefacts are different from tape saturation.

Overdrive: Can be understood as a type of harmonic distortion. This terminology is used more in production sound design. I like to use it to refer to the amount that the signal exceeds the system limit overdrive does what it says – generally pushes it further to create more pronounced distortions. Overdrive will have the effect of saturation but it will also create more artefacts that will degrade the signal. Generally, in mixing we want to drive but not overdrive the signal.

Other artefacts that may occur

It is always important to be aware that whenever we alter or distort a signal, we are likely to create artefacts that we might not like. It is always a balancing act between getting the changes we are after and mitigating the additional unfavourable artefacts that might be introduced.

Also, sometimes we use the tools precisely because of these artefacts. Understanding how a tool transforms sound is the key, then we can use it however we want. There are no wrong answers, but an awareness of the full range of impacts is important.

Smearing the transient

As we have seen soft clipping rounds off the edges. This can result in a smearing of the transient information. This might be great when you have a sound that is too edgy or harsh, but be careful that it can also go the other way and can result in a reduction of the impact of the sound.

Frequency changes

With different types of saturation as well as different amounts of drive or overdrive we will hear different amounts of alterations to the spectra.

Upper harmonic additions

This is the obvious one, we know that harmonic distortion is one of the fundamental reasons that we utilise saturation to begin with. Remember that as we drive the input more into the circuit it increases the saturation and we will pronounce the even/odd harmonics and as it is driven more, combination of both.

Low mid boost

As we are boosting the overtones, if we have a signal that contains sub-frequency, we will increase the first overtones of that signal which fall in the low and low mid region. This change can be exactly what we are after, but be aware that it can also result in muddiness of lack of mid-range clarity.

Degradation of the frequency content

As we drive in harder more of the input signal will enter the non-linear wave shaping transformation. This will increase more and more distortion that will compete with the tonal characteristics of the input signal.

Specific mechanical artefacts

Wow and flutter: These artefacts relate to the mechanical imperfections of the tape mechanism. As the speed is not perfect, the speed changes produce subtle changes to pitch. Wow is more uniform and sits in the 0–5 Hz range. Flutter is more inconsistent and occurs rapidly in the 5–50 Hz range.

Tape hiss: This is the noise floor or operating noise of a tape machine. It results from the physical and electronic properties of both the machine (play heads, etc.) and the magnetic tape. It is white noise, but is often heard more in quiet passages, or when the volume is turned up. Poor gainstaging can pronounce tape hiss.

Bias: Relates to the electrical set-up or adjustment applied to a vacuum tube that controls its operating point and subsequent behaviour and characteristics. This will impact how the tube responds when driven. This changes the saturation characteristics and the subsequent artefacts and distortions. Lo bias settings will result in a more distorted sound. Higher bias settings will give a cleaner more linear response with less saturation.

------ Pause reading here and go to Chapter 3 – FROM THEORY TO PRACTICE before reading on. Here, you will put all the information discussed in the previous pages into action exploring and playing with various saturation tools. Ultimately, you should select your tools and learn them ready to put them to use when needed in a mix. ------

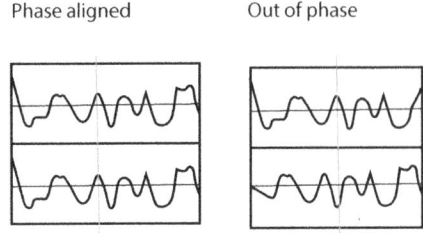

Figure 2.8 Images showing in in and out of phase

Phase

When I ask students about phase processing in mixing, most often the responses speak to the use of the 'effect' of phase from a sound design perspective. Such answers miss the power of working with this concept from a mixing perspective where we exploit our perception of space through manipulating time. We start the discussion by articulating exactly what we mean when we talk about phase processing. Then, we will present the concept of 'fixing' phase issues that may have been accidently introduced during recording or production and are causing problems in a mix. Finally, we will focus the discussion on how we use phase processing in mixing – what I believe to be the most exciting element which is utilising time adjustments to trick our brain as we build a mix.

What is phase?

Phase is the informal term given to the phenomenon of comb filtering. Comb filtering happens when a signal is duplicated and plays back slightly later.

We say that the two signals are out of phase, given that sound is periodic, it is constantly cycling, the push and pull of our speaker cones the push and pull of our ear drums as sound pressure is increased and decreased. So let us consider what happens if a signal is 180% out of phase. That means when the push happens on one track, the pull happens on another. The signal is entirely cancelled out. −1 + 1 = 0 (Figure 2.8).

Try it yourself to test out the theory! Duplicate a signal and first try to flip the phase (use a utility plugin). Also try to nudge the duplicate clip in one direction moving it later and later this will give you different stages of phase mis alignment which will change the quality of the sound played back.

If we have a more complex wave, the same situation occurs, but the relationship of the wave to itself is more complex so rather than a clean in or out of phase we will have boosts and drops at different frequencies, which leads to changes in the frequency spectrum of the sound. This is what is known as comb filtering, colloquially known as 'phase'.

What are phase issues?

We often hear people talking about phase issues or saying 'it's out of phase', or hearing 'comb filtering' in a recording or production. This refers to a misalignment of microphones capturing a single source or can also result from mis aligning samples in a production as well as effects processing which introduces time differences which cause issues. It is very important to take time to listen to and fix phase issues in the mix as it will have negative impact that will manifest as lack of clarity, lack of power and a weak centre. In the next chapter, there will be more in-depth discussion of this topic.

What are phase processes?

When we talk about audible phase processing, beginners often think of it from the perspective of sound designing technique, and one that changes the character or aesthetic of the sound. For example, if you apply a chorus effect to a Wurlitzer piano in your production to change the character or vibe of that instrument this is an example of phase processing. This is really outside the realm of pure mixing and falls into the role of the producer strictly speaking because it is a decision to change the overall aesthetic identity of the Wurlitzer sound. For the purposes of getting this conversation established, we start there, but remember that in mixing we would be careful to make such noticeable aesthetic changes.

What are some of the other phase processes? We have said already that **chorus** is a phase process, so too is **flange**, also **stereo width**, **microshift**, **doubler** and of course **phasers**.

The magic of phase

I would like to introduce a third perspective, highlighting what I believe to be the most fascinating aspect of our interaction with phase processes. This is often the part that many beginner mix engineers overlook, leading them to miss out on essential tools. It's a key approach we use when working with phase processing during mixing.

In most mixing situations, our work with phase (or perhaps it's more accurate to say time-based) processes primarily involves subtly adjusting the positioning of sounds within our virtual mix space.

A couple of examples:

- We can reposition a synth to avoid clashing with a vocal by shifting the stereo image to emphasise the sides rather than the centre
- We can also gently move a melodic high synth part to sit slightly behind the lead by delaying it just enough, either by introducing a very small delay or by using a brigade chorus to slightly blur the transient through a nearby copy of the original sound

Psychoacoustics and the precedence effect

As we develop our critical listening skills, it's important to note that some people may struggle to hear the subtle differences discussed in this book. There are inherent limits to human perception, and we can use this understanding to manipulate the listening experience. In fact, much of what we do as mixers involves this kind of manipulation. Rather than simply taking my word for it, I want to introduce you to the world of psychoacoustics and share some insight into how the brain processes sound. Whilst I will dive deeper into this topic in the final section of the book, I want to bring it up now because discussions about phase lead us directly into the realm of psychoacoustics.

One key phenomenon is the precedence effect. This occurs when two versions of the same sound are presented with a short time delay (below the listener's echo threshold). In such cases, listeners perceive a single auditory event, with the perceived location dominated by the first sound that arrives. Although the delayed sound does influence the perceived location, its effect is primarily suppressed by the initial sound. This precedence effect is the foundation for many of the 'phase' processing tools we use, such as stereo imaging, the Haas effect, chorus and phaser effects.

What is the echo threshold?

And how does listening to Steve Reich help me understand it?

At this point, I like to take a moment to listen to Steve Reich's early Phase Pieces, which are inspired by West African tribal rhythms. These pieces are fascinating because they explore this exact phenomenon. The works of Steve Reich are constructed using tape machines that play identical loops. The playback rate of the tape machines is slightly out of sync, so over time, the differences between the playback grow larger, creating a 'copy' that shifts in time. This offers intriguing insight into how the human brain organises the sounds we hear.

For example, by listening to 'It's Gonna Rain' (1969), you can experience a full range of phase processes as the time between the original and the delayed sound increases – from comb filtering to the precedence effect, chorus, flange, echo, delay and reverb.

'It's Gonna Rain' is a 15-minute piece, and you can easily find it online. Just sit back, turn it on and let it wash over you. We can discuss it after you've had a chance to listen!

Now that you've listened to the piece, let's talk about the technical aspects.

Did you notice moments when the delayed signal smeared the original sound with phasing artefacts, but at some point, it seemed to split from the original and become a distinct sound on its own? These moments illustrate the echo threshold point.

The precedence effect is generally understood to operate effectively up to 30 ms, which is considered the echo threshold for most people. This means that sounds delayed by more than 30 ms from the original are perceived as separate events or echoes. In contrast, sounds that are delayed by less than 30 ms are integrated with the original and are interpreted as contributing to the sense of size. This can affect how we perceive an object's mass, width in a stereo image or position and depth in a virtual mixing space.

What is the difference between a stereo imager/Haas effect and a chorus or phaser?

Chorus and phaser effects are based on the same principle known as the precedence effect. However, they differ in specifics regarding delay times, and the delayed signal in these effects undergoes additional modulation. Typically, both chorus and phaser effects operate within the 10-ms time delay range. Chorus effects involve multiple delayed copies of the original signal, with a low-frequency oscillator modulating the delay time between 1 Hz and 5 Hz. This modulation creates the characteristic swirling effect of the chorus. In contrast, phasing and flanging incorporate feedback into the process.

For those looking to deepen their knowledge further, I recommend Andy Farnell's excellent work, 'Designing Sound'. Whilst this text focuses primarily on sound design rather than mixing, it provides valuable insights into building the tools that mixers use, making it a helpful resource for understanding the fundamental components of these processes.

Working with phase processes when mixing is all about subtlety and attention to micro shifts in relationship between sounds. We can change the dimension of individual sounds, create perceptual movements in and between sounds to create space and separation and to build our mix environment.

------ Pause reading here and go to the next section in Chapter 3 – FROM THEORY TO PRACTICE before reading on. ------

Space processes

Many people think that picking a reverb and a delay is all about taste. I agree with that sentiment to a certain extent, and we discuss this in the section on aesthetics. There are however more details as well to consider so let's get into it. Because we often reach for a preset when working with a delay more often that with a compressor or eq it might seem as though there is not much to talk about. A poorly chosen reverb or delay can kill a mix in one action, perhaps it conflicts with the aesthetic of the track, maybe it creates messiness and muddiness, it might kill the groove or hide the vocal.

Time and space converge!

All this time talking about how we can distort the perception of space through time has our heads in a spin! This is one of the fascinating aspects of music and specifically of mixing sound that keeps me addicted to the mystery. Even though now we are talking about spatial processes, like reverbs and delays, it is important to remember that these are just an extension of the previous section – these are all still phase or time-based processes. I choose to talk of these as separate because on a practical level, in the mix, we tend to use them for distinct tasks. Please remember though the overlap!

Analog vs algorithmic vs convolution/impulse response vs AI

The art of recording, particularly the use of stereo miking techniques, was originally the primary method for creating a sense of space, depth and width in sound production. In post-production, people began sending sounds through speakers into specially designed rooms, such as custom chambers built as part of recording studios, to capture the acoustics of the space. Other types of reverb were invented, including metal plates that transmitted sound and captured the sound of the plate itself, as well as spring coils used in guitar amplifiers.

With the development of digital technology, it became possible to calculate the acoustics of specific rooms and create algorithms that applied the characteristics of reflection and diffusion to particular environments. Early digital reverbs, such as the Ursa Major, Lexicon and AMS, operated at 12 or 16 bits. To contemporary ears, these early models may sound more lo-fi compared to the high-quality digital reverbs available in today's DAWs and plugins.

As disk storage space increased in the early 2000s, convolution reverbs emerged. In this process, an impulse from a specific space (or multiple spaces) is recorded as a digital audio file and then combined with a dry sound to simulate being in that space.

The latest developments in reverb technology include AI-driven reverbs that are adaptive and can respond in real time.

Reviewing different reverb types

It is surprising to me how many people working in production don't know what a plate reverb looks like, or have a visual reference in their mind for what a reverb chamber looks like. Do a quick internet search for images and have a look if you haven't seen them yet. Once you get this visual can you see how this visual image can help you consider what type of reverb you should choose for a given task? Developing an awareness of the different types of reverbs as well as the chronological development, can help you select the right reverb/delay for the sonic aesthetic of the sound production you are working on.

Connecting to the aesthetic of the production

The space we select is so important to the feeling, mood or emotion of the track. When we listen to music, we listen to the sounds in specific spaces (i.e. where we recorded the instruments, or what reverbs and delays we use). What we are doing is by extension putting our listeners into these spaces. As listeners we inhabit the spaces of our songs, if it is a great song and an amazing production and mix, we lose our sense of reality and are transported entirely into the world of the song. How we use space has a very large role to play in this experience. We can use space to set up a sense of place, such as in a jazz club or church, or we can work with reverbs to create more subtle effects, like the feeling of being right next to someone.

Different perspectives and uses

Like working with phase processes, we can use certain types of spatial processors for different reason. We can manoeuvre our sounds in subtle ways within the mix space. We can make more significant shift moving a sound from the foreground to the mid or background. We could articulate a different space between different sounds causing differentiation. We can make a thin sound seem thicker or denser through the use of spatial processing that acts as a resonant body of the sound.

Old school still cool

Old school delays are still so popular for many reasons due to the timbre, which creates saturation that is appealing for us. For example, tape delays or tape loops are where it all began with delay and they are still a very popular and important way to work with delay in sound productions.

Parallel or insert?

Back in the day, the number of reverbs that were used in a mix were limited by what a studio could afford. When digital workstations first emerged, limitations remained based on the DSP of the computer. So once upon a time all reverbs and delays would be in parallel. There would be a limited amount of them. Nowadays, these limitations are completely gone so this has heralded in a new era of working with space processors. We can now easily choose whether we want to insert the plugin or work with it in parallel. Beginning mixers often just put the plug ins directly on each channel because they don't fully understand how to set up a parallel channel. They also don't understand that it is a choice and there are differences. Both parallel and inserting delays can be useful, but they create different spatial perceptions that is important to consider. As a mix engineer, we need to understand this and make a selection on which approach is best for the task at hand.

Some considerations when deciding on parallel or insert

- Do I want multiple sounds to be in the same space? – parallel
- Am I trying to create a sense of depth? – parallel
- Am I trying to blend or blur the sound – insert

- Am I trying to fuse multiple sounds together – parallel
- Am I trying to make a sound feel thicker – insert/parallel

Adjusting and tuning

We often combine equalisation such as low/high cut/shelf with the reverb or delay. In fact, many of these processors also include a filter section. There are two main reasons, one is about perception and the other is about mix clarity.

Perception and depth

Often we are using a reverb or a delay because we want to make the listener feel a sense of space and that they are engrossed in a sonic space that is deep and wide. A psychoacoustic phenomenon is that as sound are further away we lose perception of the high end. With a high end roll off automation we can make a sound feel as though it is moving further away. Try it!

Perception and intimacy

When we are in the doctor's office, or the train station there is not much of a sense of intimacy, there are high ceilings, harsh tiled or cement surfaces. When we are in our bedroom, there are smaller dimensions, often lots of soft fabrics that absorb lots of the high frequencies, so if we are looking to create a sense of intimacy or warmth, we might chose to roll off high frequencies to help with the feeling of being in a more intimate space.

Mix clarity

As we send a sound into a reverb or delay, we are in essence multiplying instances of this sound to play back at slightly different times. This can be problematic when the sound has lower frequency content. It can create phasing with the dry signal, as well as can add increased sound energy in the low mid region which can create a messy, muddy mix. We often roll off low frequencies and mid frequencies in the mix.

If you have certain intense that you know already, you don't want to multiply you can put an eq before the plugin to help remove these before they get to the reverb/delay. This can be particularly helpful with vocal sibilance.

How many is too many?

You can use as many reverbs and delays as you can manage, but it often leads to confusion, especially for student mixers. A mix can suffer from a lack of clarity, muddiness or an unclear aesthetic direction when there are too many effects in play. It's important to know why each effect is there and how you're using it. If you're simply experimenting and ending up with an overwhelming number of reverbs, it might be beneficial to take a step back, assess what you have and make some decisions.

From an analog perspective, there were originally limitations regarding the number of sends and returns, as well as the physical reverbs and delays available in the studio. A common technique involved using three different sizes of reverbs (for example, a short plate, a medium reverb and a large hall) along with two types of delays (like a ¼ note delay and a slap delay). Keeping this approach in mind can help you maintain control over your reverbs. Working within these loose limitations might also be helpful for you!

------ Pause reading here and go to Chapter 3 – FROM THEORY TO PRACTICE before reading on. Here, you will put all the information discussed in the previous pages into action exploring and playing with various space and phase tools. Ultimately, you should select your tools and learn them ready to put them to use when needed in a mix. ------

Meters

Meters are everywhere around us as sound makers, but it's amazing how few people have any idea what the meters do. Yes they make your studio look pro yes they somehow represent sound, but beyond that most people have no idea how to use them. They weren't made just for aesthetics; they have a very practical function. Learning how to use meters is important for many reasons. It helps support us in our listening. I see meters as my assistant and support and couldn't work without them. If you haven't learnt to read them yet, then you are missing on a powerful ally in the mix.

In the following pages, is my no-nonsense guide. We start from the familiar (the DAW) and move out into the analog world. As always, just take in what is relevant for you and come back to it when you need.

Digital meters

What meters are the default in my DAW?

The short answer is that most Digital Workstations adopted a Full Scale DB metring system. The most common meters are set to display PEAKs. Some DAW have not only a PEAK meter but also an RMS meter. Have a look at your own DAW and investigate what type of metring is the default (Figure 2.9).

Peak meters

What is a peak meter measuring?

In a modern DAW, the peak meters are sample peak meters. That means that they register the peak or high point of levels up to sample resolution. That is very fast! It gives us a

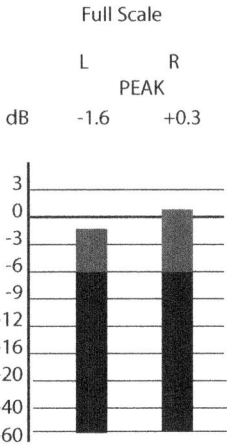

Figure 2.9 Standard meters in various DAW

very accurate understanding of whether our level is going to clip for even 1/44100 of a second!

What is the measurement scale of peak meters?

dBFS

Still confused? dB tells us we are measuring amplitude logarithmically it is the standard way that we measure loudness. The FS stands for full scale and it means that 0 represents the digital clipping point. It's the absolute loudest point on the digital system. So that means any sound that has more energy than 0 dBFS will be cut off, resulting in clipping, which is nasty distortion that ruins our beautiful music.

Why and when is a peak meters useful?

We don't want to create nasty digital clipping so this meter tells us about the loudest moments with a high degree of accuracy by making very fast moving measurements. These meters are moving faster than humans can perceive so rather than giving us a sense of loudness these meters are good for helping us to ensure that we do not reach the system maximum (if we do we create digital distortion).

When are the practical uses for this?

It is very important when **setting levels for recording** as well as for the final stage of mastering before we deliver our music to be listened to on many different machines. Anytime when we are considering how machines are interacting with the sound and need to ensure that we do not have the volume louder than the systems maximum. It is also helpful to keep one in handy when mixing but it is not as important as another meter that we talk about next. If you have set up your mix gain stage properly, you should not have to worry too much about the peak levels.

RMS meters

What does RMS mean?

It stands for Root Means Squared, but the important thing to take is that it represents average loudness.

Why do we care about that?

Its more representative of the way that we hear sound so using this as a reference point offers a more accurate indication of the correct balances.

What scale is it measured in?

Also dBFS

When is it useful? and why?

Ok so we know it's good for measuring average loudness, relates to how we hear, so the time we really start to think about this is when we are **mixing** our productions. But we also care about it when **mastering** as well

What levels should I set my mix to?

Mixing is an art and there can be a range of answers to this question. I recommend no to overthink it. Consider if your music is one that encourages loud, dynamically constrained mixes (think techno or trap, or one that encourages space dynamics (think jazz or classical). Pick a number within the ranges given.

INPUT GAINSTAGE: In contemporary digital contexts, we are mainly concerned with peak levels. Between −6 dBFS and −12 dBFS PEAK

MIXING MASTER BUS: Mainly concerned with average levels, but peak is also relevant. Between −16 dBFS and −20 dBFS RMS as well as −3 dB and −6 dB PEAK

MASTERING: Concerned equally with both measurements. Around −.1/.3 dBFS PEAK as well as −8/10 dBFS RMS

NB: these are just guides to give you some framework, feel free to define your own standards.

Analog Meters

We started off with the type of monitoring most would already exposed to because digital meters are found in our common DAW. Let's go back in time, how did they do it before computers? We will also consider the relationship between the analog world and the digital world. Finally, I'll also consider the question of whether you should actually care or not about this.

What is dBU?

As dBFS gives us a way to understand digital systems, dBU was the metering system used to represent analog systems. It is the most common measurement scale for analog systems, although other meters exist (which we will go through now). Maybe you can understand it by thinking centimetres and meters vs inches and feet. One is called metric, one is called Imperial (Figure 2.10).

What is a VU meter? What does it measure?

VU Stands for Volume Units (Figure 2.11). The 0 point on a VU meter generally relates to +4 dBU which as we relates to an average loudness dBFS level of about −16 dB.

Interacting with the multitude of tools 41

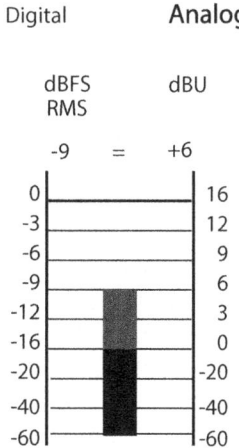

Figure 2.10 Metering scale comparing analog and digital

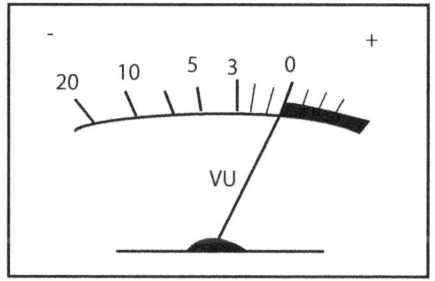

Figure 2.11 VU meter

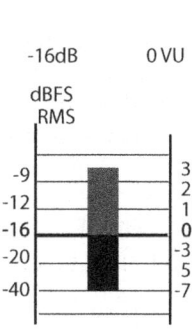

Figure 2.12 Correlation between VU and other meters

This number might sound familiar because in the previous we learnt that it is suggested above as the level that you might set as your average loudness of your mix. The response time of VU is 300 ms so it misses fast transients, actually it gives a sense of perceived loudness which therefore relates more closely to how humans hear sound. VU meters were found in many consumer level audio devices such as compressors (e.g., SSL G BUS).

Important reminder: A VU meter can be calibrated (set) to different levels, it is used as a visual cue to aid in balancing levels, it doesn't represent the full spectrum of loudness range in a signal. So although there is a standard relating to calibrating a VU in an analog context, you can and people do calibrate it do different levels. **The 0 point of the VU meter is set to the DESIRED average level.** For example, if using a plug in emulation of a VU meter when mixing in the box, you might calibrate your VU meter 0 level to –16 dBFS RMS).

Is this relevant to me?

It is not a given that the analog realm is relevant to you – here are some reflections:

- You work in Ableton and you use the combined peak and RMS and you will never work in a commercial studio – YOU PROBABLY DON'T CARE
- You work in analog studios or want to – you DEFINITELY NEED TO CARE
- You are unsure about your mixes, and the meters that come in DAW don't give you confidence – PLAY AROUND WITH VU and forget the rest of the analog systems, until you move into the world of analog
- You are just interested to understand the different way audio relates in different machines – THEN YOU ALREADY CARE! :)

New(er) formats

LUFS

What is it?   Loudness Units relative to Full Scale (also sometimes called LKFS)

Where is it used?   Initially introduced in broadcast, Film and TV to outline broadcast standards and keep the perceived volume of the different shows and adverts the same. LUFS is a further development of metring that considers both PEAK and RMS and combines this over the total duration of the song to give a simple number representation of the loudness of a single piece of music.

Why do we care about it in music?   Because it has been adopted by streaming sits such as **iTunes** and **Spotify** to regulate the level of audio on its streaming platforms.

How to read a LUFS METER
Integrated Loudness: overall loudness since the last reset
Short term Loudness: Loudness in the last 3 sec
Momentary Loudness: Loudness in the last 400ms **Loudness Range** – the dynamic range
LU – loudness units – relative to dB
True peak – same as the digital sample peak but even more accurate

How is it practically useful to me? If you are very concerned with loudness matching and being competitive on streaming sites, then many people like to work with LUFS. Many new plug ins are starting to work with LUFS meters. It has become a standard for use in mastering.

And what number am I calibrating my mix to? For starters, LUFS is mainly something that is a MASTERING consideration. The numbers given below are levels of completed music. So this is not necessarily a mixing concern.
For TV: –23 LUFS (it originally started out for this medium)
Apple music: –16 LUFS, with –18 LUFS for Spatial Audio
Spotify: –14 LUFS
Also, as we have discussed loudness can vary significantly between different genres. If you are working on a loud and dynamically constrained genre like techno, the LUFS reading can be much higher techno can be –8 LUFS (so said someone on a forum – and I'd believe it after a night in Berghain).

... a parting thought

After all this time discussing meters, I once heard a veteran mix engineer say to a student mixer *no mix engineer should think about levels/meters, just make it feel good man*. I felt that it was a bit cruel, because in a perfect world that would be great! But there is a bit more to the story. The reality is that many engineers to use meters for many different aspects of mixing regardless of skill level so it is a bit of an unhelpful statement to start off with. It is also important to remember that if you are mixing in a perfectly calibrated multimillion dollar studio with perfect acoustic treatment, then you can have more confidence to go off your gut informed by the confidence that you are hearing accurately! It is important also to consider context. The engineer that says something like this has most probably been mixing full time putting many hours of a considerable period ten years or more – they have highly advanced ears and accumulated knowledge that allows them to run off instinct. The fact is that meters are particularly helpful, they can support you in myriad ways throughout your work. If they weren't any good manufacturers would have gotten rid of them long ago. So find the tools that work for you and put them to use in every mix.

------ Read on to the next section where you will go through FROM THEORY TO PRACTICE one final time! Here, you will put all the information discussed in the previous pages into action exploring and playing with various metering tools. Ultimately, you should select your tools and learn them ready to put them to use when needed in a mix. ------

Chapter 3

From theory to practice

We have covered a lot of technical and theoretical information in the last section, but mixing is not just theoretical, it is a practical vocation. We need to turn all that theory into practical functioning knowledge that helps you approach your tools with confidence and build up your awareness of how they differ from each other in a practical sense based on you hearing the differences in action.

The purpose of this chapter is to help you make some sense of all the options out there. Often, we feel like we need to continually purchase the latest software plugins to keep up and then once we have this abundance we get lost in choice. All these aspects work against our ability to make fast and effective mix choices. You don't need to own every equaliser ever built, but you do need to carefully consider what equalisers you would like to work with. Just as a painter may choose their brushes so too must a mix engineer decide what tools to use. Now this doesn't have to be the same from mix to mix, but over time it is likely that you will likely develop a fondness for certain tools, and this will have a bearing on your sound or style as a mix engineer. What you don't want is during your mix to be confused as to which of your 100+ eq plug ins you should use. This takes us out of the creative flow, making us less alert to the mix and wastes our valuable time!

This section is designed to be tackled in a modular fashion. Hopefully you came to this page following the instructions from the previous section. After you have taken in the theoretical and analytical information about one type of processing (i.e. compression) my advice is to stop then come to this section where you can take the ideas presented above, the theory and start to put it to work in your own mixing practice. This modular approach is important. Learning every tool at once is not possible, desirable or very fun. Take the information in bite sizes, and most importantly take the time to assimilate new ideas into your PRACTICE. Without doing this step everything you have learnt will stay theoretical and you will not have effectively changed much about your existing mixing process.

This section is about making decisions, about deciding some tools you will use. Then it is about learning these tools inside and out as you explore them with your eyes but most importantly also with your ears. This is the fun stuff enjoy!

Step 1: building your mixing toolkit

We have talked about the different types and broad categories of tools that we use when mixing. So now it is time for you to make some decisions about what you want to use. Sometimes your decisions are financial, based on what you already have access to, sometimes it is aesthetic and other times it is technical. There are many great free plugins around so that you should be able to cover all of the major categories of equalisers that I have discussed within your DAW or as extra third party free plug in. If you are like me and love plugins and get excited by all of them, your job is to limit what you might use (at least for a while whilst you undergo the deep training), selecting your favourite of each category to start with. On the other hand, perhaps you have been sticking to just one equaliser and one compressor so the idea of having so many new tools is overwhelming – you don't have to add them all at once, just add them bit by bit. The important thing is not to overwhelm yourself and to go at your own pace wherever you are.

Below is a summary sheet to help you keep track and make some decisions about what tools you want to use. Enjoy!

The mixing tool kit

Thirty essential tools

Go ahead and write in below (in pencil) what you have selected as your main tool kit for the mixes in your near future. Remember this can change, but by making decisions you are starting to define your mixing approach and thus develop your own mixing aesthetic. The below is a guide of course you can use what you like!

	Tool Kit	Some Examples You Might Like to Try Out	Your Selection
	Equalisation		
1	**Digital EQ**	Fab Filter, Waves Q10, Ableton EQ Eight	
2	**Passive EQ**	Either Hardware or Software emulations: Pultec EQP1, Manley Massive Passive	
3	**Active solid state EQ**	Either Hardware or Software emulations: SSL Channel Strip, API 550, Neve 1073, Harrison	
4	**Linear phase EQ**	Fab Filter ProQ3, Waves Linear Phase, KirchHoffEQ, Weiss EQ1	
5	**Dynamic EQ**	Fab Filter ProQ3, Waves F6, Tokyo Nova, SSL DynX	
	Compression		
6	**Digital compressor**	Ableton standard compressor, Fab Filter C2, Logic Platinum Compressor	
7	**FET compressor**	1176 (many emulations!)	
8	**Opto-tube compressor**	LA2A, Tubetech	
9	**VCA compressor**	SSL Bus Compressor, API 2500, DBX160	
10	**Tube compressor**	Fairchild, Manly Vari-MU	

From theory to practice

	Other dynamics	
11	**Gate/expander**	Fab Filter G2, stock DAW gate, SSL Channel strip
12	**Limiter**	Waves L1, Brainworks bx_Limiter true peak
13	**Clipper**	Newfangled Audio Elevate, Sonnox Inflator, Brainworks bx_clipper
14	**De-esser**	Oxford, Waves Rennaisance de-esser, SSL,
15	**Multiband compressor**	Waves C6, Drawmer 1973, Fab Filter Pro MB
	Saturation/distortion	
16	**Tube saturation**	Black Box, Thermonic Culture Vulture, Sound toys decapitator,
17	**Tape saturation**	Universal Audio Studio A800, Waves Kramer Tape, Sound toys decapitator, Retro Color
18	**Transistor (solid state) distortion**	Soundtoys Decapitator, IZotope Trash
	Time and space	
19	**Stereo imaging**	Microshift, Doubler, SSL Stereo Image, Waves S1 Imager
20	**Chorus**	Brigade Chorus, stock Ableton Chorus, Dimension D, Valhalla UberMod
21	**Delay**	Sound toys Echoboy, Roland Space Echo, Valhalla Delay, SSL X Delay – too many to count really have fun with this and pick a couple of your favourites.
22	**Reverb room**	Valhalla Room, Ableton Convolution Reverb
23	**Reverb spring**	AltiVerb, Logic Space Designer, AKG B20
	Reverb plate	EMT 140 plate
24	**Reverb chamber**	UAD Capital Chambers
25	**Reverb hall/large spaces**	Lexicon, Stock DAW plugin, Altiverb Convolution
	Meters	
26	**Peak dBFS**	Your DAW has them just decide
27	**Average dBFS**	Your DAW has them! Know where to look.
28	**VU meter**	Waves VU, TP meter, Klanghelm, SSL meter
29	**Phase correlation**	SSL meter, Voxengo
30	**Graphic equaliser**	Voxengo, SSL meter

Step 2: learn your tools

Your job now is to take the time to learn these tools. Take time to focus on each type of processing individually. So that means start with equalisation rather than loading up 30 tools immediately into your project. As you are focusing on one type of processing and you have limited yourself to working with four or five different types you will be able to perform comparative analysis and learn the differences and similarities more easily. The goal is to learn these tools inside and out and to ultimately achieve a mastery of your tools. Through focused practice and experimentation, you will deepen you knowledge very quickly. Remember to do this in bite sized chunks, maximum of four or five tools at once.

A **Read the manual**
   The first stop on this road to mastery for me is always the manual. The manufacturers/developers have lovingly crafted this tool, and they have more knowledge than anyone. Often not every detail is evident by looking at the gear itself and manuals hold many secrets. Even if you have an analog emulation of famous vintage analog gear – you can still read that vintage analog manual, it will help you learn how to use it even though it is a digital emulation. These old manuals are often lovingly created and are quite amazing resources.
   TIP: Take time to understand the Pultec style attenuate and boost function in the low end, or figure out how the different phase modes of the Fab Filter ProQ impact the signal. There is a great wealth of information available from the creators.

B **Explore all the settings and listen to how they impact different sounds**
   Load up some stems or use some that I have provided (don't do this with a project you are actively mixing) and with the manual as your guide explore all the settings and listen to how they impact different sounds. Take some time to mess around with all the knobs and different settings and activate your critical listening skills to help you assimilate the style and effect of each of your tools.

C **Compare the tools against each other**
   Here, you can really notice the differences between the gear, in terms of usability and sound. Working with the same samples you loaded in on the previous step you can now a/b between them and start to make note of similarities and differences.
   Make note of things like:
       the difference in sound, how they impact the envelope and the timbre, sharpness, thickness. Try to articulate all the differences you can hear. Or make note if there are not audible differences.
       the impact on the CPU – some plugins might introduce latency, or put a large load on your computer. These are ones that you might leave for buses or even bounce down
           how easy they feel to use?
           Do they have the same parameters, are the easy and simple to work or confusing?

D **Re-creation exercises**
   I have prepared some exercises for you to go through. Find these online in the supporting website of the book. These help you learn your gear whilst also improving your critical listening skills. The name of the game here is to listen to the sound and try to remake it yourself with your own tools.

Step 3: gaining the confidence to know which tool to use when

This is the final step on your way to mastering your tools. It is not that this stage is hard, rather it just takes time. It is not something that you can rush. This chapter has been laid out with a methodology that I believe gets you to the good stuff in the fastest way possible. But it still relies on you doing the work. If you skipped the first part of this section to get to the good stuff, the summary, then you will not be able to complete step 3. You need to take the time to understand the theoretical and technical aspects of various processing tools and make some decisions as you build your own mixing tool kit, and from there you must take the time to explore and learn these tools.

Remember, as said already, this stage is one that develops over time, and only if you have good practice habits. I have laid out what I believe to be good practice habits so by following this book you will be on the right track. What we are doing is employing a **cyclical process of research, experimentation, analysis, application.**

By combining all of these elements, you will develop the confidence to reach into your tool kit and pull out the perfect tool to help you sculpt the sound to reach your aesthetic goals.

Step 4: expanding or adjusting the tool kit

When to expand or adjust your tool kit?

It is a good idea when you first start developing your approach that you limit what you use to avoid being overwhelmed. Stick with your initial choices for a while. Until you feel that you have learnt them at least. If you are happy with how they are working, Stick to them! If some or other of the tools you selected is not working for you, simply replace it. If you feel that your tool kit is missing something add it!

After you have been mixing for a while you might add a new tool to spice up or refresh yourself. If you are feeling like you want some inspiration, I always find that adding one new tool is a fun way to reinvigorate my approach. Just remember to take some time to familiarise yourself with it before you mix, learning a tool whilst mixing is probably not the best way to ensure that you are focused on the right thing which is of course getting that killer mix, not learning how to use a new tool.

Expanding your tool kit with an analog approach

If you are fortunate to work in a commercial analog studio, you will notice that there is a great deal of outboard processing available and that often include in this is a great deal of equalisers. Some people say why? When there are console EQ's on every channel on the mixing desk?

The answer is that engineers and producers, often use these tools to create and shape the colour and the aesthetic of the music. They might like to give the vocals the sheen and brilliance of the Neve 1073 high shelf, but want the power and compactness of the SSL low end. Maybe they want the gutsy low mid of the Pultec on the bass with the smoothness of the API on the guitar.

So whether we have an analog studio or not, we can also take this approach and overtime build our equalisation tool kit to include a variety of different 'colours' and approaches to equalisation that will suit a variety of use cases when we are mixing.

POP quiz: am I a master of my tools?

Before moving on from our in-depth study on mixing tools and processing take this pop quiz. If the answers to the below questions flow pretty easy – congratulations it is time to move onto the next area of study. If these questions are still challenging to answer, then it reflects that more time is needed studying your tools before you are ready to improve your mixing.

Equalisation

1. Give an example of when would you use the Pultec Style attenuation shelf?
2. Why is the passive style EQ great for using on mix buses, and the low end?
3. What is the slope of a Neve style Lo cut? Why does it appear to beef up the low end, even though it is cutting it?
4. What is the difference between an active and passive equaliser?
5. How can an eq help with signal that is out of phase with another? i.e. snare top/bottom
6. Which type of equaliser will be really good for removing ringing resonances? Which will not be helpful and why?
7. What part of the circuit in analog equalisers introduces the famous 'colouration'?
8. When is it useful to use a dynamic eq? How is a dynamic EQ different from multi band compression? How might you use these differently?

Compression, dynamics and envelope shaping

1. What are the characteristics of the different compressor circuit types?
2. What is the role that time plays when working with compression?
3. If you are using presets what is the one element you must adjust manually every time?
4. What does the knee of a compressor do?
5. What is the difference between a gate and a downward expander?
6. What controls on dynamic processors impact the envelope of the sound?
7. What is the difference between a dynamic EQ and a multiband compressor
8. What is side chaining? How do you set it up?

Saturation

1. What is the difference between soft clipping and hard clipping? Where do we find hard clipping? Where does soft clipping occur?
2. Name different types of saturation?
3. What is added through the process of saturation?
4. What is the natural harmonic series?
5. What is 2nd-order and 3rd-order harmonic?
6. List some of the artefacts that can be introduced when using saturation
7. Name some of the changes that can occur to the frequency of a signal after applying saturation
8. Why is distortion not a very clear term to use?

Phase processes

1. What is the difference between using phase processes in production and phase in mixing?
2. What is an example of a phase issue that might occur during recording?
3. What audio processing tools are in the category of 'phase processes'?
4. What is the relationship between space and time?
5. What is the precedence effect?
6. What is the echo threshold in ms? Explain the phenomenon
7. What is the difference between chorus, phase and flange?
8. How does Steve Reich fit into all of this?

Space

1. What are the different types of reverbs? How do their characteristics differ?
2. Why do old school digital reverbs and delays sound 'lo-fi'?
3. What is the analog approach to setting up reverbs and delays?
4. Why would you use a reverb as insert over parallel?
5. Why is using reverbs in parallel the most common way to apply them?
6. What is the relationship with depth and frequency?
7. How can you play with perceptions with reverbs?
8. What is tuning a reverb and why do you do it?

Metering

1. What meter is standard in your DAW?
2. What is the difference between peak and rms meters?
3. When do you use a peak or rms meter?
4. What levels do you set peak and rms to when mixing?
5. What is the reference system for analog?
6. What is the relationship between digital and analog metering? i.e. calibration equivalent
7. What is a VU meter, and what is it useful for? What does the 0 point represent?
8. What does LUFS stand for? How is this different and similar to the previous meters discussed. When is it used?

Chapter 4

Developing advanced ear sensitivity

if you can't hear it you won't fix it

You can have all the tools in the world, have all the latest plugins and still be a bad mixer. The most important tool we have as mix engineer is our ear and our brain. But we need to learn how to apply them. This section should be viewed as your ear training boot camp. It is recommended to complete small amounts of this every day, each day gradually building up and challenging yourself further. But it is also a great resource to return to as often as you need to. I like to return to ear training for multiple reasons, even though I have already undergone extensive training. Sometimes if I am sick, or tired, or maybe I have come back from holidays and might need to calibrate again before recommencing work.

I offer you my training regime. The instructions and overview here tell you what you will find on the accompanying website/app where you will find audio examples for you to practice with. You can supplement it with one any online/app based tools as well. Find what works for you, but make sure that you do it!

Note on practice

- Say out loud the frequency you think it is, that way you can't cheat. Knowing if your instincts are correct is just as vital as knowing if they are wrong. Only by knowing your weak spots will you actually improve
- Move through these quickly and intuitively
- Take note of the frequencies you get confused and take time between tests to practice these
- You build up over time quite quickly so the key thing is to 15 minutes regularly – if you can do it every day, or even a couple of times per week, you will notice an improvement quickly!

Exercise explanation

Frequency recognition

The frequency balance of our mix is so important, this is one of the most important skills to have as a mixer. Just as a jazz musician must be able to hear different chords and scale formations, we need to be able to identify pitch as frequency so that we can rectify issues make decisions on what elements of sounds we want to mould, bring out or remove.

We start with the sine tone for its relative ease of detection, then we complexify it somewhat by introducing noise that is filtered to the band and finally we apply filter to more complex material that we will find in our own mixes (individual stems, groups and master).

Amplitude recognition

Being able to hear changes in level is important for multiple elements within the mix. Firstly, we use this as we create balances with the faders. It is used as we do our gainstaging. We also employ this vital tool when maintaining unity as we process our sounds for example when compressing. It is also employed when we create panning and consider the balance over the stereo field. We also utilise this skill as we create depth in our mixes and position sounds further away (combination of amplitude, diffusion and frequency)

Compression, saturation, clipping

Detecting different forms of processing is an important skill in particular when we are analysing other music. It helps us to draw out favourable effects in mixes that we love, or references we are given from clients. It also helps us to accurately identify issues within mixes that we are working on.

You will be asked to recognise different types of waveshaping and subsequent artefacts that can occur such as: Soft clipping, Hard clipping, Compression, Saturation, Even harmonic distortion, Odd harmonic distortion, Wow, flutter, Transient smearing

Space and phase

The perception of space is one that can be so tricky because it is a psychoacoustic effect. In stereo mixing, we are tricking the mind into believing that we are in a space when it is all happening in our minds. It is so fascinating to think about! But we need to be confident as mixers to identify these subtle perception shifts. We might need to identify phase issues occurring in our mix. We might want to identify a specific effect in a reference, we might also need to identify issues in production and consider better options.

We employ ear training to:

1   Hear a range of delays and interpret their style rate and amount of repetitions
2   Determine a range of different reverberation types – plate, spring, room
3   Determine the decay level of various reverbs
4   Determine chorus, flange, phase effects on sounds
5   Hear comb filtering of different lengths

Ear training boot camp

Please refer to the **online resources** to engage with this section:
https://www.janearnison.com/mixing-in-flow-online-resources
Whilst you can find your way through these exercises as you like, it can be helpful to be given a training regime. Below is my advice on how to step through these exercises.

Go through these at your own pace. Do them regularly. Do not move on until you feel confident with the each preceding stage. Overwhelming yourself is not helpful. Take it easy, have fun with it! Also, remember that if you have not worked for a while, or you are feeling foggy or sick, you can use these to help focus your mind and your ears, so they are good to come back to throughout your career, not just as a learning tool when you are beginning.

Suggestion: Do these 15 minutes in the morning 5 times per week, for 2 weeks - then review your improvement

Beginner

Week 1

1   Frequency recognition: Sine tone: recognising 8va separated frequencies

Week 2

1   Continue with week 1 activity
2   Amplitude recognition: recognise 6 db changes in a range of sound examples
3   Space and phase: name the delays, name the reverbs

Week 3

1   Increase difficulty by changing sine tone to filtered noise – recognising 8va separated frequencies
2   Continue with amplitude recognition 6 db
3   Name the phase effect: chorus, flange, phase, Haas, slap echo

Week 4

1   Continue with week 3 frequency exercises both sine and white noise 8va bands
2   Amplitude recognition: recognise ±6 db and ±3 db changes in a range of sound examples
3   Space: determine delay length

Intermediate exercises

Week 5

1   Sine tone: recognising 3rd 8va separated frequencies
2   Continue with amplitude 6 dB and 3 dB
3   Determine different types of wave shaping

Week 6

1 Increase difficulty by changing sine tone to filtered noise 3rd 8va bands
2 Amplitude recognise: ±6 db and ±3 db, ±1.5 dB changes in a range of sound examples
3 Space and phase: determine decay level in reverbs, estimate delay length of short examples

Advanced exercises

Week 7

1 Frequency: identify cuts and boost in musical examples 8va bands
2 Amplitude: determine changes to amplitude of specific sounds within arrangements
3 Space and phase: detect comb filtering

Week 8

1 Frequency: identify cuts and boost in musical examples 3rd 8va bands
2 Amplitude: determine changes to amplitude of specific sounds within arrangements
3 Space and phase: detect comb filtering of different lengths
4 Detect different types of wave shaping

Chapter 5

Setting up your listening environment

Optimising the listening environment is one of the most common challenges that people encounter. Despite all of the dedication and training that we have undertaken so far in this book, you might walk into your bedroom studio try to perform a mix and still be making the wrong decisions because you are not hearing the mix accurately. It can be disheartening to know that there are factors more-or-less out of our control that can greatly impact our ability to mix well. So what is going on? The two culprits are the monitors you use as well as the room that you work in (which is actually even more significant).

There are two major realities which create barriers to solving these problems, which often sit on opposite poles. The first one is the vast amount of specialised knowledge that one must possess on acoustics and psychoacoustics to be able to make assessments and changes themselves. The second pole is the vast amount of money that is often spent setting up mixing rooms. It often seems that you either have loads of money to throw at hiring specialists to do the work, or you need to possess the knowledge yourself. Is it as simple and dramatic as this? Does it simply mean that if you do not have the money and you do not have the knowledge that you are doomed? My opinion is that the 'vastness' of these poles can be reduced greatly and it can be a gradual process of investment and acquisition of both knowledge and treatment over time as you develop into your career. So in other words, start where you are and accept that it will be a process of learning and investment and building up yourself and your studio.

If you are just starting out the first question might be – can I still mix effectively in substandard conditions? The answer is yes, but with some caveats. I have mixed in every condition imaginable; with headphones, in bedrooms, on planes, in the best commercial studios in the world and in a range of semi-treated studios of my own making. The one clear line is this: I can create good mixes in all circumstances **but it is a lot easier, faster and more enjoyable** when I'm confident that what I'm hearing is what I'm hearing! i.e. when I mix with reliable monitoring and a treated room. So whilst I am glad that I now have a reliable working environment, it didn't start out that way and I still managed to do some pretty ace mixes. The key to me is always confidence. **Am I confident in my decisions because I am confident in what I am hearing?**

First of all we are going to look at ways that you can start to make decisions about your monitoring choices, then we will look at different stages of developing your room and finally important methods to double check when you have to work in a room that is 'untrustworthy'.

Monitoring

The first step is monitoring, because we can't do anything until we can hear our music through speakers. So first question you are confronted with is which monitors should I use? Ask three mix engineers and they will give you three different answers. I am constantly asked to recommend monitors to people. My answer is that you should buy the best monitors that you can afford. The reality is that monitors these days are all very good, the technology has been refined and the parts cheaper. If they are all good, which ones to use – how do I pick? The best way to figure this out is to listen to them. Listen at school studios, friends studios and homes and go into music stores. Whilst you can do online research and read armchair experts wax lyrical about why one monitor is better than the other, ultimately it is a personal decision, and *you* only know what you like when you listen for yourself. There have been great mixes performed on a wide range of monitors, from cheap monitors like Rockit to expensive monitors like APC. Don't over think it – find your budget range, **listen**, go with your gut.

Stick with your monitoring

The next most important thing to note is that once you pick your monitoring, it is a pretty good idea to stick with it for a while. This is because another way that you can manage deficits is to learn the sound profile. Companies like Adam Audio have three-tiered system that allows you to grow with the sound as your experience and budget grows, this way you maintain recognition of the sound profile you are used to but get increasing clarity as you invest more.

Can I or should I use headphones?

There is an increasing trend of headphone monitoring for mixing, so let's address it. The reality is that if you have a properly treated room mixing with speakers is advisable to use them! Mixing in a room is better for many reasons. One is that it is less taxing on your ears and the likelihood for ear damage is less. Another is that mixing in headphones can pronounce certain things (left right panning) and reduce others (depth perception). There is the issue of the phantom image, and there are a range of other psychoacoustic anomalies. However, for many people starting out, and not having a devoted mix room, using headphones is an attractive and realistic option. So my guide for when to use or not use headphones is this – if you cannot spend over 400euro on your monitoring you should consider headphones as a first step. You can get a really good quality pair of headphones for around 400euro, whereas your monitors for 400euro per pair, would not be a quality considered suitable for mixing. Headphones of course also solves the problem of the room.

Some argue that most music is made for headphones these days so we should listen and mix in headphones. Just because music is made for a format is doesn't necessarily mean that it is the way that we should monitor when mixing. It is true we should listen and check the translation in these formats. We might use the same example of listening in a car, or even listening on a laptop. Mixing is trying to create a final output that translates well on multiple formats – radios, cars, different headphones, home stereo, high-quality PA, laptop speakers, etc. So we want to hear all the possible details. This is why specialist monitors for mix engineers in conjunction with a flat response room are still the best solution.

As we move into spatial audio that utilises object based approaches, more and more engineers are utilising headphones for mixing in spatial contexts. There are certainly different considerations here, and binaural mixing is certainly performed optimally on headphones. But again, if you want the option for spatial multi-speaker listening, then you should have access to a studio that has multiple speaker set-ups.

The room

If you are going to spend upwards of 1000euro on your monitors, then it means you are getting more serious about mixing and wanting to hear more accurately. So, it is also time to start considering your room. When first encountering this area, it feels very complex and can be hard to make sense of a lot of the information on the topic of acoustics. However, if you take it gently, you can gradually build up your knowledge. If you cannot afford to hire an acoustician, you can start to learn the basics and make practical physical adjustments to your room to improve your mixing experience.

In order to get the most out of the following discussion, it will be helpful to also to deepen your research around acoustics. Look to the expanded reading bibliography at the end of this book as well as searching for resources that cover the below key concepts.

Key concepts to explore are:

- First reflection path
- Room modes and anti-modes
- Speaker boundary interference
- Absorption vs diffusion

Room analysis tools

AMRAC Room mode Calculator – easy to use, not a real time calculation but based on the room dimensions, Can play modes and listen to the suggested issues.
REW Wizard – free, complex to use
Sonarworks – paid, easy to use

Making alterations to your room

In the following section, we go through different common scenarios that developing mixers are faced with. These are organised in increasing 'expertise' or professionalism. The idea is to help you optimise as much as possible whilst being realistic about your situation.

Bedroom mixing

If you have to work from your bedroom or another room in your house, there will be limitations with the positioning, there will also be other furniture and items that you have to contend with. You also need to be able to live in this space and for it to feel like a home so subtle treatments will work best. Let's look at what you can do to optimise this space.

Selecting monitors appropriate for the room

- Massive monitors with sub low end are probably not a good idea, and definitely not a sub-speaker, because this will amplify potential problem areas that you cannot effectively treat
- Front ported monitors are important if your desk is against a wall
- Have your headphones on standby for mix checking

Setting up the room

- Bookcases are great to have behind you when mixing because they act as diffusers
- Decoupling your speakers (i.e. putting foam/rubber underneath them) will help with sympathetic resonances from the desk
- If your speakers are against a wall, you will need to consider speaker boundary interference – by moving them forward and backwards by small increments you can ascertain if the low frequencies are boosted or cut
- Try to position you speakers so that the listening position is equidistant from the side walls, to ensure the first reflection path is balanced
- You might be able to make a baffle to act as a temporary wall to help with the balancing of sides
- When deciding where the best place is for the mixing station – put on your reference list and walk around the room, also try to place the speakers in different parts of the room. Where does the reference track sound most 'correct' to you? That is the spot to pick
- It is a good idea to do a test on the room once you have decided on your positioning. You can use various software or online tools for this, see above list

Commencing treatment in a multi-purpose space

If you are able to do some treatment to a space but it remains multi-purpose you will likely have limitations with how extensive the treatment can be. You should observe all the points already raised above, but perhaps in this space you can be a bit more selective than you might be able to in a bedroom. The first step in treatment is often the first reflection path. The second step would be to start to introduce some bass trapping.

First reflection path

Treat the first reflection path with absorption and/or diffusion (Figure 5.1). You can start with simple foam or invest in more significant panels that also offer bass-trapping/low-frequency absorption (see below for more details). If you want it to fit with the aesthetic of your multi-purpose room, you will need to consider that as well. Curtains also offer absorption, but no significant in low frequencies. Consider the impact of windows as well and where they are positioned relative to the first reflection path they can create flutter echo as the reflections bounce off the shiny reflective surface. The back path is often well treated by placing a bookshelf behind the listening position.

Bass trapping

Understanding acoustics can be challenging, and without this knowledge, you may fall victim to misleading products on the market. It often requires some trial and error.

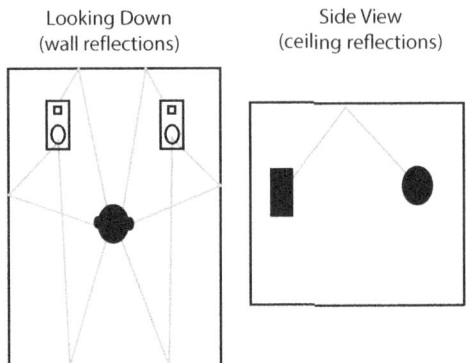

Figure 5.1 First reflection path

One important point to keep in mind is that if you need to address low frequencies or sub-bass, thickness is essential. Therefore, if you come across thin, inexpensive panels claiming to treat frequencies below 100 Hz, it's likely that those claims are exaggerated. Another key concept is mass. In a small room, using around eight panels can yield decent results, but using twelve would offer even better performance. However, overcrowding the space may become an issue. In such cases, corner bass traps can provide a subtle yet effective solution.

Ceiling and floor

Ceiling clouds go above the listening position. If you have a hung ceiling, you could put some rockwool inside that for a similar effect. If you are working in a room where you can't make holes in the ceiling, this can be a problem. Some people create a solution by working with trusses. The floor shouldn't be forgotten, a great rug can do wonders to a reflective surface and it has the bonus of helping create a nice vibe in the room.

Doing self-treatment in your devoted mix studio

Once you are in your own studio, you can optimise the space specifically for mixing without needing to compromise.

First, take the time to find the ideal listening position. Set up your monitors in the room and listen to how they sound. You can adjust their positions slightly and trust your instincts to find the best-sounding location. If you're already in your room, move one speaker around and note how the sound changes in different locations. Confirm whether your current positioning is the best. If it sounds better elsewhere, you should rearrange your room, as this is the fastest way to achieve optimal sound.

At this stage, you will also become aware of the room modes in an empty space. Using an AMRAC mode calculator will help you understand where resonant energy is most active. Learning to use Room EQ Wizard is beneficial at this point, but you might also try simpler tools like Sonarworks to achieve usable results. One is free, while the other is paid; choose which suits you best.

Consider your options for diffusion and absorption. If you're handy with woodworking, you might consider making your own treatment solutions. Alternatively, you could explore room-in-room construction or purchase competitively priced room treatment options. Your choices will largely depend on your skill level and budget.

A key concept to understand regarding room modes is that you will likely need a significant amount of bass trapping and absorption to control the mode spectrum in your room. For instance, in my studio, which is 36 square meters, I have 20 panels designed to absorb low-frequency modes, in addition to having extra absorption in my hung ceiling. If, like me, you have budget constraints, you can take a modular approach, making improvements as your finances allow. It took me about two years to achieve a studio set-up I was happy with. While it would have been ideal to have it professionally treated from the start, I couldn't afford it. This process helped me refine my ears to better interpret the room acoustics, so I feel I now know my room intimately.

Generally, the first step in treatment is to address the first reflection paths. If you have significant low-frequency issues, it's advisable to use panels that also act as bass traps (i.e., they should be thick!). After addressing the first reflection paths, consider treating the corners or angles of the room, where low-frequency buildup often occurs.

To gain a more accurate understanding of where to place additional absorption, utilise software like Room EQ Wizard to visualise potential modal hotspots. You can also walk around your room to listen for areas of sound buildup (modes) or dropouts (anti-modes). Pairing this listening approach with the AMRAC Mode Calculator can be both informative and engaging.

Once you have positioned your speakers and listening area, and applied treatment to your room, the final step is to use EQ for subtle adjustments to both the frequency spectrum and the time delay between the speakers. This is especially important for multi-speaker set-ups. Many modern speakers now come with DSP options for this final treatment, as do audio interfaces, so consider these features before purchasing monitors.

Now it's time to listen and acclimate to the room. You may want to tweak the set-up for a while as you get used to it. Eventually, you will feel the room settle, and you should stop making adjustments to stabilise your listening environment.

Summary treatment for a devoted mix studio

- Take time to locate the ideal speaker positioning in room
- Position the monitors relative to the back wall considering SBI
- Be sure to treat the first reflection path
- Additional treatment of room modes (bass trapping) in corners and hot spots
- Using software to perfect timing issues and eq balance: Sonarworks or EQ Wizard
- Take time to listen to references, and to learn the room – you might also like to do ear training in the room and see how sharp your ears are once it is treated

Chapter 6

The power of references

References and a personalised reference playlist are very powerful tools for a mixing engineer. References help us at every step of the way from setting up our studio, to training our ears, they help us when we need support and they help us ensure that we are doing what the client wants. So it is worth devoting a section to spell out their importance.

A vital tool for mix engineers

- Using references to learn a room
- Using references to select the right headphones or monitors for you
- Using references if you do not trust your listening environment
- Using references if you don't trust yourself (i.e. sick or tired)
- Using references to train your ear sensitivity and critical analysis skills
- Using references to learn how the greats do it
- Using references to make sure you and the client are on the same page
- Using references to check that your mix decisions are good and your mix stands up against the industry

Why don't aspiring engineers use them?

You can see that there are so many uses for references, yet, beginning mix engineers do not use them. There are a few reasons as far as I can see for why not.

Firstly, they feel like it is cheating, or that they want to be original and not copy. If you look at any tradition of learning a craft, one of the most important steps for the apprentice is first to copy and learn from the master. Only once they have mastered the technique are they then encouraged to expand and forge their own style. This is true for painters, sculptors and musicians, and also it is true for mix engineers.

Ok so now we have that out of the way, let's look at another more critical reason that developing mixers don't use them. **They do not know how to.** This is something that needs to be learned like everything else. Using references requires critical listening skill – so if you haven't yet completed the critical listening training, then you need to go back and develop your listening skills first. Once you have critical listening skills in place, we can start to consider references, how to select them and then how to use them efficiently.

Building a reference playlist

A personalised reference playlist is a list of songs that you as a mix engineer have developed over your mixing career. They may build up over time, and your list may change a bit, but there should be at least one or two songs that have remained in the list from the beginning – these become your touch stones when you are in unfamiliar territory.

Songs you know well and love: The reference list is examples of songs that you know very well, so that when you play them in different rooms, on different gear you can interpret the colouration of the room on the song that you know so well.

Songs that have been mixed excellently: Some pinnacle songs that you think are great, but also are held to a high standard by the industry. You can look at past grammy winning engineers, as well as look at top mix engineers sharing their favourite mixes. It is all out there online. The songs that have been mixed excellently do not necessarily have to be songs that you love personally, or that fit your general style. But focus in on the mixing!

Top songs in the genre/s that you often mix in: These songs may be the ones that change, with you on different mix jobs, or perhaps as you grow over time and change as a mix engineer. Hip hop is mixed differently to acoustic folk, so make sure you have key songs in genre that you mix in.

Songs for specific sonic features: There are some songs that feature extremely low frequencies, that are great for testing the low end of speaker clarity. There are songs that detail transient clarity, and other songs that can reveal the clarity of depth and the size of the mix space your room/monitors offer.

Uses for references when mixing

When you are mixing for a client, or even if you are mixing and have specific mix goals in mind, it is often required that you can perform comparative analysis between your mix and a reference track. The goal here is not to copy the track. Of course, the production is different so it cannot be exactly the same. There are a range of uses for comparative analysis when mixing; here are some of the common ones:

Balancing the relationship between the kick and the bass – this varies wildly from genre to genre, and depending on how well you room is treated, you might be getting a false impression of the bass response in your room – if you listen to a reference, you can compare the bass as it sounds in your room to your mix and make better decisions for your mix.

Finding the right level for the vocal in relation to the other sounds – again this is one that varies a great deal across genres. It can also be a very personal thing – so when an artist gives me a mix and says they love the way the vocals sound, I'm also considering how the vocals sit in the mix relative to other instruments.

The transient material, and how aggressive it is – whether it is trap style high hats, sibilant ss, rhythmic arpeggiation or picking sound from a bass guitar – attacking transients can cause ear fatigue quickly. When you have ear fatigue, if you have been listening to the same thing repeatedly, the brain protects itself from over stimulation. Transients are the loudest parts of the song that can be attacking our ears when we are paying attention to other aspects of the sound. Some people become hypersensitive, and others become dull to it. Either way it means bad mix choices. By checking your mix with your reference and paying attention to the transients, it will recalibrate your ears to ensure that you are getting those transients sitting correctly.

Depth perception – how far can you trace that reverb tail? Can you hear the dimensions of the room and really get a sense of the sound that inhabit the foreground, mid and background. To be able to discern the zones of your mix and have clarity between these is the mark of a great mix, being able to hear the end of particular sounds in space is a good marker if your mix is going well. Having a track in your reference playlist that does it well is great to keep you alert to the challenge.

The clarity in the mid-range – this is an area which is another common problem area, and although it might be something that you consider when balancing the kick and bass, depending on the song the issues may be other instruments. You might think your mix is clear, because you have been listening to it for ages, and your ears have accustomed themselves to really focussed listening, it is not until you listen to a top-quality reference mix, that you hear that your mix could be a lot clearer and that there is a diffuse 'muddiness' to your mix.

There are of course more details that you might well be interested in a particular mix; however, these pointers hopefully get you started paying attention in the right way when listening to your references and comparing them to your mix. We will pay closer attention to these issues (and more!) and how to fix them in your mix in Part Three of the book when we look at common mixing problems and how to solve them.

Exercise: mixing playlist

Start your own mixing playlist. Include one of your favourite songs that you love the sound of. Include a song from a mix engineer that you admire. If you like you can research reference lists of big engineers and consider adding a third song from a great mixer.

Setting up references when mixing

We have talked a great deal now about references, but one of the most critically important factors has not yet been raised. If you do not set up your references properly as comparative listening they are rendered basically useless. First of all, I am going to explain how to do it properly and, secondly, I am going to justify why setting them up properly is **critical** to using them effectively.

Loudness matching

We should know now the basic principle that the louder something is the better it sounds to us, so it makes sense that we need to ensure we do not have this bias influencing us when comparing.

How to match them? – simply turn down the fader of your reference mix.

Match them using a meter or using your ears? It is up to you. I tend to go for my ears, because when they are meter matched, the reference will generally sound louder because it is limited (more on that below).

Echoic memory – the importance of the quick switch

This is a BIG one so don't skip it. A button click is better than a mouse click which can be time consuming. Take a moment to set this up and it will help you so much in the future. The reason it needs to be quick and automatic relates to a psychoacoustic concept called Echoic Memory.

Echoic memory

This is a psychoacoustic concept that explores how long we can retain detailed information about sound. It typically lasts up to 3 seconds. Which means that after that time the details of the sound in our memory are lost and our brain starts to reinvent the sound changing the quality. We can train our ability to hold on to detail longer as it shifts into the Auditory Working Memory; however, if we focus on other tasks or hear other sounds, all of these also impact our ability to remember accurately as our mind loses focus. You can see where this is going! It is imperative to set up your references properly, otherwise, the a/b that you do will give you false or misleading information because your memory of the sound has already changed by the time you play the second sound.

Reference should bypass mix processing

NB: When using digital audio files that you drop into your DAW project – make sure your reference mixes do not pass through the master bus processing.

If you get used to setting up a mix bus summing channel in your DAW, this will make it easy to add additional tracks that pass directly to the output. If you work in Ableton, you can send the reference track group out to the external outs bypassing the master.

Chapter 7

Workflow

We have spent a significant amount of time in this book exploring the tools you use and considering how your brain interacts with sound through critical listening. Much of this thought process is very specific and focused on small details. However, if we approach a mix with this level of detail, we risk getting lost, prolonging the mixing process indefinitely and ultimately losing our sense of direction. Many students express their inability to finish mixes and their struggles with lack of confidence in their choices. Developing confidence and the ability to complete mixes relies on establishing a reliable workflow.

This section examines an area that is often lacking among aspiring mixers: a dependable workflow. Without it, finding consistency, focus and confidence will be challenging.

We approach this by talking about concepts that emerged from psychology. These perspectives will help explain how having a workflow can enhance your attention and analytical engagement with the material you are working on. Additionally, they act as a guide to help you establish methods that can help you ensure that you are working in a productive and responsive way. A structured approach makes mixing more enjoyable, which builds confidence and generally leads to better-sounding mixes. This is the core aim of the book: to help you achieve a state of flow in your mixing process!

Mixing in flow

Mixing in flow is so fundamental to my creative and professional practice that I named my book after it. So let's dig in and learn about flow.

One of the most renowned studies on flow states is by Mihaly Csikszentmihalyi, who published a book titled The Psychology of Optimal Experience in the early nineties. According to Csikszentmihalyi, the crucial aspect of an optimal experience is that it serves as an end in itself. So there must be a sense of enjoyment and fulfilment through the act of doing regardless of the outcome. One of the benefits of experiencing flow states, beyond enjoyment, is that those are moments when you make your best decisions and are most likely to achieve successful outcomes. If you're interested in exploring the science further, I highly suggest looking up this book. However, for now, let us reflect on the insights regarding mixing sound.

How to get into flow states when mixing?

If we review the common characteristics, there are some very strong hints at what is required in order to achieve flow states when we are mixing.

Skills

The first point relates to having the required skills. Skill acquisition is of course the first stage in any task. **But what skills exactly?** In my time teaching mixing and mentoring producers and mixers, one thing is very clear to me everyone can mix, or can demonstrate the appearance of being able to mix…but not everyone can mix well.

The first, and most basic step is to understand what mixing is and what essential tasks a mix engineer performs. The second step is to be able to perform those tasks well (this is the hard bit!) That means you need the understanding and the ability to action that understanding when given a 'sonic problem' to solve. In other words, it is not just enough to have a rough idea that when mixing you compress a signal using a compressor and that in order to compress a signal you move the knobs on a compressor. In order to achieve a successful flow state the skill needs to be in the ability to critically listen to the sound, consider an approach to using a compressor and then be able to execute that approach and get the desired result.

Workflow

The second point relates to workflow, which can be described as a "goal-directed, rule-bound action system". The first part, having a process, refers to this "rule-bound action system". In the context of mixing, it means establishing a workflow that you consistently follow for every mix. This includes the order in which you complete tasks, how you organise your DAW files, your naming conventions and so on. The key here is consistency. It's less about what specific actions you take and more about making a deliberate decision about your process and sticking to it. While each mix engineer may have a slightly different approach, we all share one common trait: consistency in our methods.

However, the story doesn't end there. This is an aspect that many beginning mix engineers often overlook. It's not enough to simply have a process; you also need to include goal setting and actionable steps. In other words, when you receive a mix project, the first step should be to listen to and analyse the tracks, make assessments and create a plan for what needs to be done. Combining goal setting with a thoughtful and consistent mixing workflow will set you on the path towards a state of flow.

NB: Regarding the skills related to achieving flow, it's important to recognise that it exists on a continuum. Over the years, I've found that as my skills improve, I am able to enter flow states more quickly and maintain them for longer periods. It's not the case that you can only reach flow states after 20 years of experience; rather, the more you practice, the better you become at it. This process involves building neural pathways associated with good habits, allowing you to find your groove with greater ease.

Distractions that kill flow states

Once you have acquired the skills in audio processing and in critical listening, have developed a strong mixing workflow which includes analysis and goal setting it should be automatic flow states?! Almost, but we are not quite there yet. We need to deal with one of the most common and persistent killers of flow states – distractions, so what are those distractions?

The biggest killer has to be distractions from our communication devices. So that means, phones should be turned off or put to do not disturb and very importantly, put out of site, not on the desk next to you upside down – there is some interesting research that shows that this is still distracting. Emails should be closed, all notifications on the computer you are using to mix should be turned off. All other programs that are not related to mixing should be closed.

The second distraction is a poorly organised mix project. So this means, get your naming clear and making sense for you, get the colour coding set up. I recommend deciding on a colour coding scheme and sticking with it. It doesn't matter whether your drums are blue or red, but it does matter that once you decide on the colour you stick to it. This helps you to get into flow faster when you don't have to ask yourself which colour the drums are.

How do I know when I'm in a flow state?

Your concentration is so intense that there is no attention left over to think about anything irrelevant, or to worry about problems. Thus, your self-consciousness disappears (which means that you trust your decisions and don't second guess yourself). Your sense of time becomes distorted (if you have had times when you are working and you feel almost like you blinked, and three hours went by…then it's likely you were in a flow state!). This sounds like a great place to be! I'm sure you have all experienced it.

If we can identify with these feelings, even if they are fleeting, it means that we can train ourselves and create opportunities for these feeling to be maintained for longer. We feel good and our mixing output is better. Much of this book is geared around supporting you to develop the skills and habits that will encourage and support you experiencing flow states.

Embodiment

Embodiment is a much talked about term in contemporary psychology and wellbeing contexts. It relates to the way that we are present and how we observe and listen. To be embodied literally means to be in our bodies – i.e. cognisant of our body, to be aware and receptive to our physical senses. In contemporary life, where we live much of our lives through computers, and screens and connect virtually, we tend to live much of our lives from the neck up. We are lost in thoughts and often loose the connection with our bodies. Have you ever been working on your computer at your desk and after a while realise you were sitting in a strange position and have a leg ache or neck ache? How often are you scrolling on your phone and realise that you just lost 30 minutes or more? Descartes famously said 'I think therefore I am', and Freudian Psychoanalysis seems to encourage us to honour the thoughts over all else and though these ideas persist in popular culture they have actually been debunked and better theories and concepts have taken their place. New theories and concepts all include awareness and investigation of the body. Without the body we miss many important messages, experiences and signals. To be embodied, to be present and in our bodies is better for our health but it also makes us more aware of our environment, and able to evaluate our responses to various situations.

How does embodiment relate to mixing?

Beyond making us more alert and focused, being embodied directly relates to our experience of sound. On a base level sound is vibration, legendary mix engineer Dave Pensado recently said on Instagram 'Sound is s physical thing, it's a vibration that goes through our body, it's not just something that you hear, it's something that you feel'.

If we are not connected and interested in how sounds make us feel physiologically, then we are missing out on important information. It can help tell us how a sound is sitting in the mix, whether the transients are too intense, where there are any strange low frequency clashes. It will tell us if the groove is hitting right, it will tell us if the emotion of the music is effective.

It also can tell you if you are too stressed out and not listening with the right attitude. It can remind us that we need to take a break. Remembering that breaks are important for ear health as well as focus and good decision making.

How do I become embodied?

To become embodied there are simple exercises that we can do. If you have ever meditated, you will be familiar with these types of exercises. You can adopt some suggestions below of find your own method, the main thing to note is to take a moment to stop, think about your body and consciously *feel* your senses. All of your senses. Taste and smell are perhaps less important in mixing contexts but they can be powerful tools to bring yourself into the present moment and can be used to reset yourself on a break.

Exercise 1: Simple breathing

Take five deep breaths through the nose. Count slowly to five on the inhale and five on the exhale.

Whilst you are breathing try to feel the air moving through your nostrils. Don't think about anything else but that sensation.

Exercise 2: Releasing breath and tension

Take a deep breath in through the nose. Hold the breath for three counts. Release for five counts. As you release feel your whole body loosen. You might like to repeat this several times.

Exercise 3: Simple movement and breathing

Take a walk around your room focusing on your breath. I find this one especially helpful if I have been sitting working at my desk for a few hours.

Exercise 5: Gentle awareness listening to the environment

Take five minutes to try and notice every sound in your environment. Once you hear it, name it and let it go. Listen as openly and curiously as possible and enjoy the sounds as they arrive at you.

If you are working in a soundproof room, you might want to leave this room. It is a great moment to go outside for a 5 minute break to reset your ears.

When to practice these techniques?

1. **At the beginning of a session:** The tendency to arrive at your desk, open up your computer and just start working is very high. But by taking a moment to consider

your plan, to focus your attention, to bring awareness to your body can help set you up for a very productive session.

2   **Whenever you realise you have lost the connection:** Practising embodiment is advisable anytime you realise that you have been disconnected. If you have any body pain from sitting in strange positions, this is a clear signal that you have left your body! It doesn't take long, as little and five breaths can do it.

Build simple embodiment interventions into your workflow, the habit will help you mix better and enjoy yourself more whilst you are doing it.

Establishing a workflow

After all the talking of concepts, some of you are probably hungry for some simple guides or examples. Whilst I will not do this too much in the book, because I believe it is important for you to establish your own approach, when it comes to workflow it can be nice to compare notes. This also includes discussion around the hot topic of knowing when the mix is actually finished. You will see that with a reliable workflow this question can be easily answered.

A guide to approaching a mix that you actually finish

1   Start the session with a moment of embodiment (see above), be clear about your goals for the day
2   Listening and analysis (taking notes) of the track you are mixing, centre your ear with references
3   Mix approach and plan (you might do some additional research here to make sure you are armed with the right approach and knowledge to get the best results)
4   Mix set up – including gain stage, mix bus and setting up any key analog processes
5   Fixing problems (now you start going through your checklist from your analysis notes)
6   Optimising the mix – balance, relationships between parts, fits genre etc. (here the reference track is very important)
7   Improve the sound 'make it expensive' (in this moment really check in with the pre-mix and make sure your mix sounds better than what you started with)
8   Consider the whole song – progression through the song – is it exciting, does each part lead to the next, is there movement, how do the transitions hit? – do you want to play it again when it is finished? (From here on play key transitions and play the song from start to finish.) Automations, flow and movement between parts and between sections

Important tips

- Remember to set up your reference at the beginning so it is very easy to switch between the mix and the references
- Don't forget to use the references THROUGHOUT – at every stage of the process
- Decide on your levels at the beginning
- Decide on mix bus processing at the beginning

- When you are listening and planning your mix – make detailed notes so you have a check list of things that you need to do, it's also the time when your ears are fresh and receptive to what works and doesn't.

When is the mix finished?

- When all your check list items are checked off
- When you feel that the relationship between the reference is in the right place (obviously it won't sound 100% the same – that's not the point!)
- When you have slept on the mix and come back fresh the next day and you like it
- When you send the mix to the client and they like it:)

Chapter 8

The pre-mix preparation

Set it up right and make life easy

DAW or desk set up

Whether you are working in the box, analog or hybrid, you need to take a moment to set up your mix. This is not anything revolutionary, but it is worth going through what you should be aware of. This is something that I used to not ever talk about but over the years I have noticed that it is something that is not necessarily obvious and certainly is something that not everyone does.

Set the sample rate and bit depth

Generally you will find out what the stems you are mixing are and just set up your project to the same settings. However if you find that someone has exported stems at 16 bit, I would recommend to set your project at 24 bit minimum. For those of you wanting to work with Dolby Atmos, it is important to note that you will need to work in 48 kHz so if the project is at 44.1, you will need to convert the stems.

Import stems organise name colour

Again some people feel quite strongly about this. Either that it is a waste of time talking about it, or that it is one of the most important elements. I am not here to tell you what you need to think, but rather I'm here presenting information so that you can make your own choice and develop your own workflow. I would be part of the latter group that feel that this is an important part of the process. For me and many others, this is an important step that helps to make it easy to know where everything is. This is another process that helps you get to the sounds quicker and move between them with more ease. This means more immediate response and more flow when mixing.

Consider a system that works for you and stick to it! Creating a habit around this organisational approach helps for muscle memory and speed when working.

Set up your meters

In the previous section, we talked in detail about different approaches to metering. Metering is really important for me. I like to utilise visual aids as well as my ears. I have an

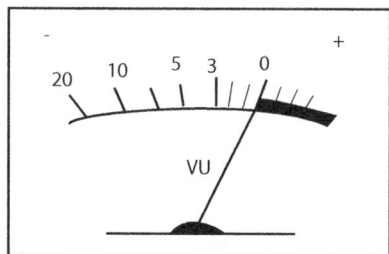

Figure 8.1 VU meter on the master bus showing the dial at 0 for the most part of the mix means gain staging is complete

old school analog background so I like to work with VU meters (see Figure 8.1). You can and should explore meters and find something that works for you. Understand how to read it and use it. We set up the meters in the beginning because you need to refer to them in the next step which is gain staging.

Gain staging

After I have imported and organised my stems, I then perform gain staging. It is important to note that I do this before engaging in routing for reasons I am about to explain! Firstly, let's just get clear about gain staging. It is a much talked about and also much ignored part of the process. I constantly see mix projects with poor gain staging and students seem confused by what it is and what it isn't. In my opinion, gain staging should happen quickly, it should not be over thought. It is a quick check and a simple adjustment to levels before you start your mix. If you spend too much time gain staging, it means you have probably begun the mix which we don't want.

Why gain stage?

Traditionally, gain staging was essential to ensure that mechanical and analog systems operated at their optimum levels. If gain staging was not done correctly while working with tape and an analog mixing desk, there was a risk of having levels that were too low. This could lead to a disproportionate relationship between the noise floor and the mix level. Conversely, if the levels were too high, excessive distortion and saturation would occur due to overdriving the equipment.

In analog processes, the concept of unity is important. This refers to the optimal operating level. In analog systems, the 0 point on the meters typically represented this optimum level, or unity.

Some argue that gain staging is unnecessary in digital contexts because the noise floor issue is no longer a concern. However, if you use analog emulation plugins, these systems have been designed to respond similarly as analog gear. Therefore, maintaining levels relative to unity remains important for controlling how the unit responds.

Even when working entirely in a digital environment without any interest in analog emulation, I still believe that gain staging is crucial. The visual representation we get from mixing provides important insights into how the components are interacting. For instance, if you are driving too hard into your master, you might miss out on identifying frequency masking issues that will be shown in a VU meter on a properly gain staged mix bus.

How to gain stage

If you are still reading I imagine, I have convinced you of the benefit of gain staging. Now the simple part is how to do it. Decide on your loudness goals for the mix. For example, I generally use −18 or 16 dB RMS.

Now look at you master bus meter. If your mix is hitting around −18 dB, already you don't need to do anything. If it is louder/quieter, you need to make some considerations how much louder it is. I like to make adjustments based on the calculation of 6 dB and its relationship to double the perceived loudness. I work in multiples or divisions relating to that 6 dB point. If everything is just a bit loud, then I'll turn it all down 1.5 dB if it is quite a lot over the −18 dB point then I will turn everything down 6 dB. If it is quieter than −18 dB, then I do the same thing but turn everything up this time.

Why adjust all stems at once? More often these days, there has been careful consideration about the balances of the project during the production stage. Whoever has been working on the track (whether it is you or someone else) has spent time crafting the internal balances of the parts. I find that it is more effective for me to start with this at the beginning of the mix rather than creating a blank slate. I try to honour this approach unless of course I am given a blank slate and asked to reimagine the whole mix. By trying to maintain the existing balances from the beginning, I am also more conscious about what I change from the original, which allows me to be able to be conscious of the development of the mix from the production, rather than it just being another interpretation.

How to adjust all stems at once? This depends on your DAW, but you can generally select all the stems and adjust the clip gain of multiple stems. You can also insert a trim plugin as your first element in your DAW.

If you are working analog, you can easily manually adjust the trim pot on your channels. Easy and fun.

Individual stem issues? After you have calibrated the stem levels you might find that there are one or two sounds that take up a lot of energy. That is, when they sound the meter goes wild. In this instance, I would take a closer look and listen carefully to see if it is disproportionately louder in the context of the mix. If it is, then I will make an individual adjustment of that signal so that it is more balanced with the rest of the sounds in the mix. If it is not, then it tells me that there is frequency masking occurring, which I can fix in the mix.

Maintaining this level throughout. I adjust it until my level meter gives me a reading that my average loudness is hitting my goals. This level I then maintain throughout my mix.

Routing

Many engineers have specific approaches to routing. For example, Andrew Scheps is known for his famous parallel compression and back bus technique, while Michal Brauer utilises a method called 'Brauerisation'. Exploring these techniques can be a fruitful

Mixing into Subgroups

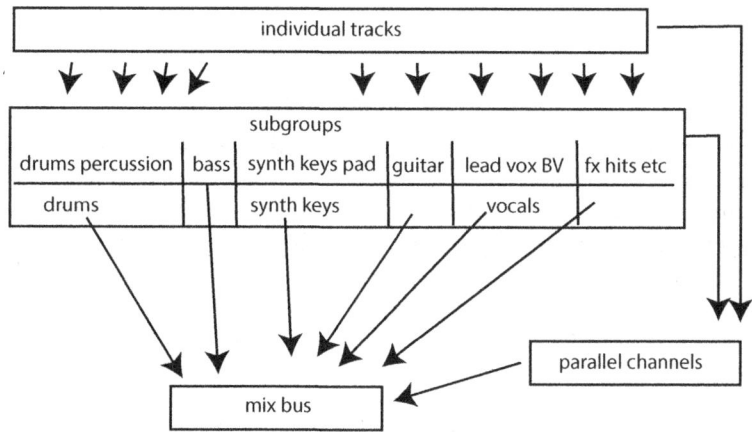

Figure 8.2 Routing example for a typical mix

source of creativity for a mixer. I recommend that as you start out, you keep your approach simple and enjoy the opportunity to delve into these techniques later as you gain more knowledge.

A straightforward configuration for routing often involves creating subgroups inspired by practical considerations, such as bouncing stems after the mix is completed.

It's also crucial to decide when to apply processing to groups of tracks rather than processing them individually. For instance, using VCA (Voltage Controlled Amplifier) compression on a drum subgroup can help the individual sounds feel like a cohesive ensemble. Similarly, background vocal harmonies can benefit from this approach to fuse the layers into harmonious textures (Figure 8.2).

Set up your references

For full information on references, refer to the section discussing references. The most important thing to remember at this point is to set up your references CORRECTLY. This means:

1 Make sure you loudness match them with your mix levels now that you have gain staged
2 Set them up so you can almost instantaneously switch between the reference and your mix. This means a one button press ideally. This relates to the concept of **echoic memory** discussed previously

Setting up the mix bus

If you work with a mix bus now is the time to set up your mix bus before you begin the mix. Remember we mix into the processing here rather than put it on at the end.

If you haven't used a mix bus yet, don't start now. Go through the book and when we get to Part Four the discussion of mix bus processing will be taken up in detail. You can then incorporate it into your mixing.

Note on workflow: direction through the mix

The next part of the book presents a problem-based approach to mixing, by focusing on specific issues or desires that we might identify during the critical listening process. It offers ideas to help you interpret what you are hearing and then consider various ways that you might go about responding to what you heard. This is not a step-by-step guide for mixing. So, before we get into the specifics let's take a final moment to conceptualise a general approach through the mix that you might take.

Common approaches

1. Bottom up – starting from the foundation elements, building up the arrangement to the vocals
2. Starting with the lead elements and building the arrangement around them
3. So-called top down mixing that focusses more on broad strokes on groups – mix bus and subgroups
4. Problem based – which considers only the issues that are noted in the critical listening session
5. Fix it as you hear it – approach where you critical listen and follow your instincts and respond to the music fixing issues as you hear them

When you first start mixing, it is probably easiest to work on a **bottom up**, where you can build the mix picture in a progressive fashion. This works well when you are still developing your critical listening skills. It will also work better for certain genres – for example music with strong drums and bass elements will thrive on this approach. There are potential traps with this style that need to be considered. There can be a tendency with doing a bottom up type mix is that you make big changes to the existing production. This might be a good thing, but it might also create problems if you have a pretty solid production pre-mix that you want to keep the vibe and just make adjustments.

> If you have a solid pre-mix, you might find that working with problem based or 'fix it as you hear it' – which favour your critical listening skills identifying areas that the mix can be improved and adjusting only what needs it. These approaches are more advanced and relies on your having developed highly sensitive and analytical listening. So often when you are starting mixing it might not be the most ideal way of progressing, because perhaps when you listen to the mix overall you are not fully grasping the issues.

Checklist:

How do you know when you are ready to begin mixing?

Mixing training

- ☑ I have **researched** the tools I am using and developed an in-depth knowledge of the concepts and technical information
- ☑ I have **linked this theoretical knowledge with my practice** through doing practice exercise experimenting and exploring these tools (without the stress of completing a mix, but for the joy of learning the process and the gear itself!)
- ☑ I have practiced and refined my **critical listening** and can determine between different frequencies, different loudness, spatial positioning, etc.
- ☑ I have a **plan** for how I will proceed through the mix (a mixing workflow)
- ☑ I have **set up my room for optimal listening**, I have spent time learning my rooms idiosyncrasies

Mixing preparations

- ☑ I have performed a **first listening and analysis** of the pre-mix which has given me points to achieve in my mix
- ☑ I have **set up my session** and organised my mix space
- ☑ I have **removed distractions** that could kill flow states – turned off my phone and closed-down other apps, etc.

Part Three

Creative problem solving – inside the mix

Summary

In the third part of the book, bolstered by the in-depth training of the earlier parts of the book, we find ourselves inside mixes. Every mix will present itself with challenges. This section approaches common mixing challenges titled as if it were feedback from a client or note from critical listening. We identify the problems through listening and then deep dive on a range of potential solutions to these problems. This is the fun of the creative and subjective components of mixing! Not every one of these ideas will be relevant to all mixes, sometimes just one is applicable, other times multiple of these questions and problems are valid in a single mixing project. It is up to each mixer to analyse the mix and determine what points are relevant. There are often multiple solutions to a sonic problem and depending on the sound and the aesthetic of the production, different solutions might be interesting. In this section, we see the four elements of mixing starting to come together. Critical listening is at the forefront, analysis and interpretation follow, considering the range of technical approaches, and finally decision-making after assessing our options and aligning with the sonic aesthetic of the music we are working with.

Chapter 9

Not everything can be front and centre

What are we hearing when we are listening?

- The mix surface is flat – there is no real depth
- The instruments don't feel like they are working together they feel separate and not an ensemble of sounds
- In more extreme situations, particularly with dance music and pop music where lots of dynamic control might have been already used in the production process – the mix might not feel like it can breathe, it often might be experiences as a flat plane that is happening in front of us, we do not feel immersed in the sounds
- There might be a bit of irritation as all sounds are in some way shouting over each other to take the front position
- The mix often will have no dynamic range, or sense of space

Concepts discussed in this section

- Building a mix picture
- Dimensions of listening
- Initial balances
- Playing with perception

Organising and prioritising

Music is often called organised sound. Whether you are a classical musician scoring an orchestra, an electro-acoustic sound artist or a pop music songwriter there is a process of prioritising the roles and placement of the parts, whether consciously or otherwise. Sometimes it is the conventions of the genre that dictate it. For example, vocal pop music will always have vocals as the priority out loud and proud in the front. In certain classical music forms, the goal is for total balance and evenness so that parts are blended together, in more complex forms of experimental electronic music the focal point may be constantly metamorphosing throughout the piece.

When we consider this challenge in the context of mixing, we need to become aware of the decisions that may have already been made throughout the piece's development. Sometimes these decisions will already be clearly made and no other adjustments are needed other than to ensure that we honour these decisions. (Please refer to my gain-staging section about how to honour these decisions.) This is also a good time for listening to reference tracks and consider the wider aesthetic of sounds that the piece you are mixing relates to.

Building a mix picture

Building a mix picture is often a great place to start. Some people take this process very literally and like to draw out a visual representation of their plan. If this might be you, refer to the David Gibson visual mixing book *The Art of Mixing* to help you gain inspiration.

Whether you think visually or not, it is important to make decisions about the priority of sounds in space. We are creating a perception of a 3D experience for the listener, the more we are clear on where the sounds should be within this virtual mix space from the beginning the higher the changes of creating an impressive mix experience. This relates not only to the size of the mix but also to the ability for the listener to be 'inside' the mix. I find that in all great mixes the listener has a sense of being able to move between and around the sounds.

The dimensions of listening

Sound has always been of vital importance for our survival. Over our evolution have developed keen abilities to determine positions of sound in space. Initially, it was to help us detect if there was something dangerous approaching, or perhaps to track our dinner when we were hunting. Today, we are happily not required so often to need to utilise this skill in this way. This ability is something that we can (and do) play with as mixers. These capacities are understood as psychoacoustics, which is the study of how humans perceive sounds. (You will notice throughout the book that this word comes up. Gaining an awareness of how psychoacoustic concepts work in everyday life and how much of our mixing plays with how humans perceive sound is very important to mastering your craft as a mixer).

So what are the dimensions of listening? These are the same as the dimensions of space, because sounds always exist in space, whether it be a virtual creation or a physical phenomenon. Depending on the type of mixing you are doing these will be slightly differently organised. We will focus for now on stereo mixing (but keep in mind that spatial/immersive approaches to mixing have slightly different sensibilities. This is discussed in more detail in Part Four of the book)

Stereo mix

Width – left and right, stereo image
Depth – Foreground, Mid, Background
Height – Frequency, transient
These three dimensions are always at play with each other as well as the 4th dimension which is TIME.

Setting initial balances

We can set up the mix very simply by having a clear idea of our priorities for the different sounds in the production. Once we have built a mix picture or made a list of most important sounds, we can start to create some initial balances across the dimensions of the mix. You might like to create initial balances by making adjustments to the faders and applying simple panning to create the first placement of the sounds in the mix. This may also have been partially done by the previous producers/composers before you start mixing, or you might be starting from a clean slate. Listening in mono can also be helpful so consider whether stereo imaging is doing the bulk of the work in defining depth, and if so make some more conscious choices for the depth field as well as the stereo field. Even though at this point it is not likely to do any equalisation, it is a good moment to consider the vertical/frequency arrangement of the mix and note any conflicts or adjustments that might need to be made.

Playing with our perceptions

Once we understand that organised sound is more than just putting sounds together, but that sounds organised in time also takes on a spatial dimension, we start to become more aware of the concept of psychoacoustics. Let us deal with the three dimension and consider the technical and practical applications that mixers employ.

Width

Width is created through panning. Width is also created through playing with our perception of space by playing with the time relationships between the left and right, also known as phase processes. These such as stereo imaging and chorus, phasor etc. It is one of the moments where mixing allows us to come to face with the nature of reality itself by touching on the time-space continuum.

Depth

Depth is created through simple adjustments of the faders. It is also created through adjustments to frequency – psychoacoustics tells us that as sounds get further away from us humans perceive less high frequency. We can also create a sense of depth by manipulating the perceptions of sounds in reverberant space. For example, if we imagine a sound occurring far away from us in a large space like a warehouse, we are likely to hear the reverberation of that sound rather than hearing the direct sound. Thus we can create these types of virtual situations in our mix in order to create a sense that a sound is far away from the listening position.

Height

Height is created through frequency. Our ability to perceive height is the least developed ability that we have. We learn this particularly when starting to work in immersive or spatial mixing format. However, there are ways that we can create a perception of sound rising or falling in physical space and frequency is always involved.

Have you ever heard of a Shepard's tone?

This started out as a technique employed by avant-garde composers such as Risset and Tenney but it has since found a home in popular genres of EDM Dubstep. This format of music often features a breakdown which builds to a big drop – the Shepards tone is a great tool for creating a never-ending anticipation. It does this through creating a sense that sound is constantly rising higher and higher. If you are interested to find out how to make this effect, there are many places to help you online. In the context of mixing I mention this to establish a sonic reference

Time, perception and relativity!

As I've already hinted at, when we consider time in the context of music everything starts to shift and bend! Our perceptions of space are actually created through adjusting the time of the repetitions of a single sound source. As we include time in our awareness we realise that at every moment in a sonic work the individual sounds are changing their characteristics; whether their frequency, their amplitude or their envelope. In the context of our perception of space, we also need to understand that we create a sense of space relative to other sounds.

By making conscious decisions about the priority of sounds in the production we can make adjustments to the position of all sounds relative to each other in the mix space. This results in a clearer mix that is deep and wide and immerses the listener.

Chapter 10

Clean it up

When I talk to beginners about what mixing is, they will say something like 'mixing is about cleaning up and optimising the mix' – and they would be right! It is indeed an important part of mixing. So, let us pay attention to this to break down exactly what 'cleaning it up' means.

What are we hearing when we are listening?
- Individual sounds that don't sound like you expect them to, i.e. they sound unnatural (maybe there was a problem with the recording process)
- Individual resonances (again likely from issues with recording, but also can be introduced during production)
- There might be 'artefacts' such as clicks and pops

Concepts discussed in this section
- Frequency
- Clicks and pops

Frequency

Optimising individual sounds

The initial element that people are likely to focus on would be the obvious issues. Sometimes there can be a problem with a particular sound that is audible. For example, the lead vocal might often need individual attention to deal with artefacts such as sibilance, or there might be a weird sound happening in the guitar part. It obviously makes sense to fix anything that you can hear doesn't sound right.

To solo or not to solo

Warning: sounds do not exist in isolation, and so be careful to listen in context

The approach for optimising individual sounds is something that people argue about a bit. The question is whether to solo and adjust or not. Some people contend that soloing is to be avoided because all sounds are heard relative to each other. When you solo, you hear the sound differently and then make adjustments on the soloed sound that are not optimised for listening to the sound in context. I agree that soloing too much is dangerous, but I do still do it during my mix. I always work quickly when I solo, keeping in mind the limits of echoic memory (discussed on the references section) I perform lots of comparative analysis to ensure that the solo works in context. This means turning on and off the processing when out of solo to interpret the results amongst other parts.

Some mixers also like to approach the idea of optimising the lead sounds only and then building the mix around that sound. For example, if you are working on a vocal pop track, you might start with the lead vocal in solo, You will make this vocal sound as amazing as possible on its own, and then as you add the other elements ensure that this amazing vocal remains so as you add other elements. The adjustments will occur on the other elements in relation to how they negatively impact the lead vocal.

To sweep or not to sweep

There are many sources that suggest you can find the frequency in question by sweeping through the frequency range with a notch boost, generally at about 10 dB of gain. I am not a fan of this approach. The main reason relates to the concept of echoic memory that I have already talked about. The brain cannot recall an accurate memory of the default sound for a long time and that as soon as we hear other sounds this confuses the brain. When we sweep through the frequency range to find the problem frequency that we heard, we actually quickly lose track of that specific sound and often start to prioritise other things that the boost sweep enhances.

If you are relying on sweeping, it means that you have not yet developed your critical ears enough. It means that it is probably a good idea that you revisit Part Two of this book and in particular the section on critical listening and find the frequency exercises. Those mix engineers that sweep all the time are also not really developing their ear sensitivity and critical engagement with sound as frequency.

A compromise

I know that identifying the particular problem frequency can be really tough, and it feels a bit mean and also unrealistic to say to a beginner, just practice until you can pinpoint the frequency. My suggestion then is to do the following.

- Listen to the sound and try to describe the issue. Thus helping you to focus in on it (e.g., the kick drum sounds loose)
- Give your best guess for the frequency range of this problem. (e.g., 300 Hz). You will learn this by studying beginner books that help to match description of sounds with specific frequencies.
- Go to that frequency range and then if you need to perform localised sweep to identify the exact sweet spot for the issue. Remember to a/b and to move quickly
- Cut the area that you believe to be the issue
- Listen in the song and a/b, analyse the change and determine whether you solved the issue
- Repeat until the exact issue is solved

The above method still develops your analytical, and critical ears, whilst allowing that you might need some help to pin point. With this method, overtime you will increase your ear sensitivity and hopefully need to rely on sweeping less and less.

Remember to practice! If you ever feel that you need to go back to Part Two – do it! This is not a linear process. Continue to refine and fine tune your own mastery.

Accountable EQ or turn it off

In an ideal world, if we would achieve such high-quality recordings that little to no equalisation would be necessary. Conversely, if you are building your music using sounds from a commercial sound pack, you would expect those sounds to be of good quality. If you are skilled at combining those sounds effectively, you would hopefully avoid introducing new issues, thereby reducing the need for extensive equalisation. However, the reality is sometimes recording is not masterfully done – perhaps the equipment is of lower quality, or the arrangement has one or two issues. Or perhaps the recording was created in a generic fashion and the mix needs to assert the sonic aesthetic. In these cases, as mix engineers, we find ourselves needing to employ equalisation. We always approach processing with purpose, and **if that purpose is not met it is imperative that we turn it off.**

I often notice that there is a lot of copycat eqing where students copy a curve that they saw in an online tutorial. While some of these may work well, many do not, as they often miss the crucial step of listening and applying EQ specifically to the sounds. I always go through each decision with my students, discussing their choices and A/B testing the settings. We talk about what they were trying to achieve with their particular equalisation. If we can't hear the intended change, we then explore in what ways the sound was actually altered. Often, these changes aren't desirable and the student was not paying attention to this. The key point here is to engage the most important tool we have – our ears – and to critically assess the sound. I recall a lesson from an orchestration teacher, who reviewed one of my scores. He pointed out a specific note and asked, 'Why is this here?' I had no idea why I included that particular B-flat in the clarinet part. This experience highlighted that I was indiscriminate in my choices and lacked focus and direction in my compositions. I carry this principle with me into all aspects of my work, especially mixing and equalisation, and I now share it with my students. If you can justify a wild EQ shape, go for it! But if you cannot articulate why it's there and can't recall the goal you were trying to achieve with it, then turn it off! This indicates that you are not truly listening or making informed adjustments, which likely means your mix is lacking direction.

Wild EQ curves and phase cancellation

I often notice that student mixers tend to use wild and excessive equalisation settings. Your crazy equalisation is making your mix weak (check out the more power sections for a detailed attention to phase issues and comb filtering). More on this topic when we get to the section abut power, because yes you guessed it phase cancellation has the sonic result of weakening the sound.

Critical listening

You will notice in this section I did not give you specific frequencies to boost or to cut. There are many resources that provide these 'cheat codes' if you want. I encourage you to use these if they help you. I remind you to refer and engage with the critical listening exercises in Part Two of this book. Engage in your critical listening, analyse the parts and make notes on what is working, what sounds good, what is getting lost and go through each of these points with a focus on frequency recognition.

What type of equaliser should I use?

A quality digital equaliser is a great tool for this type of cleaning equalisation. You want something that is flexible that will easily give you the results you want. There are no hard and fast rules here, if you get an idea that a particular EQ might work well on the sound, trust your instincts and try it out. It might be an idea to have an EQ that can offer linear phase. Armed with your mixing tool-kit you will start to remember the way the different equalisers affect sound and quickly make intuitive decisions on what is appropriate.

Clicks and pops

Where do they come from?

Adjusting clicks and pops is something that many people don't often consider. Depending on the style of mixing you are doing, you might pay more or less attention to this aspect. However, there are several reasons why it is important to address this issue in contemporary mixing practices.

Firstly, much of today's music is recorded in home studios with budget equipment and untrained engineers. In the past, music was typically recorded to a higher standard. The advent of computers and affordable recording gear has made it possible for anyone to record anywhere, which is a fantastic development, but it has also affected the quality of audio that mixers find themselves working with.

Secondly, the aesthetic and general style of music production in the mainstream have shifted towards a more 'perfect' sound. Whether or not you are working within the realm of this particular type of popular music is irrelevant; what happens in the mainstream influences everything. The ripple effect of techniques and processes flows from the experimental to the popular – and vice versa.

Finally, as digital audio resolution improves and our processing tools become more advanced, we tend to hear more detail. This means we pick up on the desirable elements of sound as well as the undesirable ones. If you've ever worked with tape processing, you know that running audio through tape can help smooth out any artefacts. Older technologies didn't have the same issues, thanks to the nature of their reproduction methods.

How do I fix them?

This is a perfect time for an AI tool. My favourite is the iZotope RX de clicker. Also useful is mouth de-noiser for vocals. If you don't have an AI tool, you can employ your trusty noise gate. This requires a fast and transparent gate, it also requires more skill than an AI tool. You will have to set it and listen to ensure that you are only removing what you want and not taking away parts of the sound that you want to keep in. If you are clearing up a vocal track, you will likely perform some type of De-essing. This will be discussed in full in the later section where we focus on the vocals.

Chapter 11

Don't get stuck in the mud

The previous section already considered cleaning things up and this section continues where that left off. We might already notice an overlap. This section focusses more specifically on the low and low mid frequency area where we have problems with 'muddiness' and looks at a wider range of approaches and solutions to these issues.

What are we hearing when we are listening?

- Sounds when combined feel 'muddy' – this is characterised by having a lot of energy in the low mid and low region 100–500 Hz
- Sounds are not very well separated, they are feeling like they are in a soup with difficulty to place individual sounds
- The mix is very 'centred'
- The mix has no depth

Concepts discussed in this section

- Frequency register and arrangement
- Frequency masking
- Repositioning in space
- Side chaining

Frequency, register and arrangement

Developing a mental picture of the notes as frequency

Let us consider the frequency range. Middle C at a 440-Hz tuning is 261 Hz (approx). Many people are surprised of the actual pitch of 260 Hz when they hear it. The common perception of this single tone upon hearing it is that it is a higher value. When I first started mixing I was surprised at how low the fundamental frequency was in many instrumental parts. As you develop your critical listening through ear training, you learn to position the fundamentals more accurately. Hopefully, it has a knock on effect to your arranging technique if you are also a songwriter/composer. It is important to be clear that the fundamental of most signals lie in the low-frequency band between 100 Hz and 300 Hz. No wonder that many arrangements have problems with so-called mud when

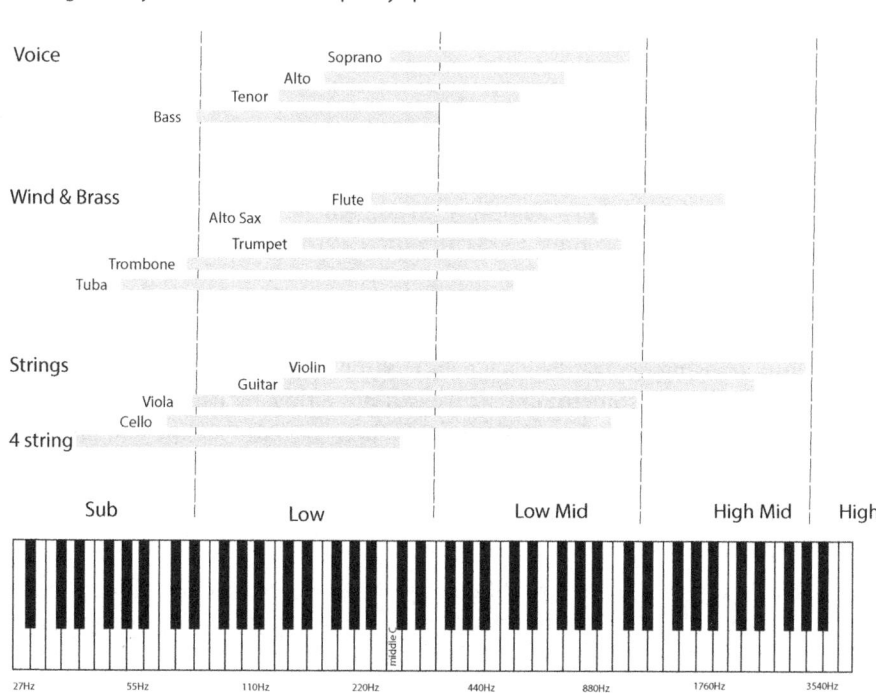

Figure 11.1 Frequency range diagram including instrumental registers

multiple parts are playing the same note in the same register (Figure 11.1). This situation is further complexified when we consider the full spectra of a sound and the way the harmonics of each sound will interact with each other.

Stacking bass instruments

We can add into the mix the pre-occupation in contemporary music with bass and sub frequencies. These days we might not have only one kick and a bass, but three kicks and two basses that we somehow have to balance in the low register. (We will get into more detail on that later in Low End Theory.)

Decisions need to be made!

One of the culprits for a poorly executed production is lack of decision making. This will also lead to a poor mix if the decisions are not made during the mix. For more information about this, please refer back to the section not everything can be front/centre. If these decisions haven't been made in production through the arrangement then in order that the mix works, they must be made now. Not everything can be heard in the same area at the same time.

So now that we know the cause of the common problem of muddy and bottom heavy mixes, let us look to ways that we can fix them in the mix.

Adjusting register

Once we start to get a mental picture of the register of various instrumental choices and understand where the fundamental pitches are it can help in a few ways. I will mention how these can be fixes in the production or arrangement stage, as well as how we can approach them from the same angle but with different tools in the mix.

Octave displacement

You might think of using octave displacement in your arranging before it gets to the mix.

But in the mix it is also something that we can do, either by using a pitch adjustment harmoniser type tool, or through equalisation where we cut out the fundamental and focus on the first overtone.

Chord voicing

You might decide to voice the chord differently across different parts. This is not something we can explicitly alter in the mix, but we can bring out harmonic resonances through harmonic distortion (think tube and tape saturation).

Frequency masking

We talked in an earlier section about frequency masking. This is particularly important to revisit in this section. The mud in question often occurs because too many sounds are in the same register all trying to be heard at once, Not all of them are actually being heard as they are being masked in part or in whole by other elements. The result is a kind of source bonding, and the result is a muddy sound that has no individuation and shape.

There is a great deal of discussion about frequency masking. There are many new tools, in particular AI tools, that can help identify it. Perhaps you develop a technique where you utilise these tools to detect and irradiate the masking. Remember through that mixing is a delicate process, and sometimes you want the masking and other times you don't! When you employ a tool as always ensure that you are in control not letting the tool run wild! You need to be making decisions as always about what part of which sound you want to place where in the frequency range (Figure 11.2). Each of these decision is building the mix.

Give and take

Often to support the balances, we want to approach it like a give and take. If we reduce the fundamental from a sound which lets the same frequency resound stronger in another instrument, then we might like to boost the 1st overtone in the sound that lost the fundamental so its relative loudness remains the same. Then, the two parts combined will result in a thicker combined sound.

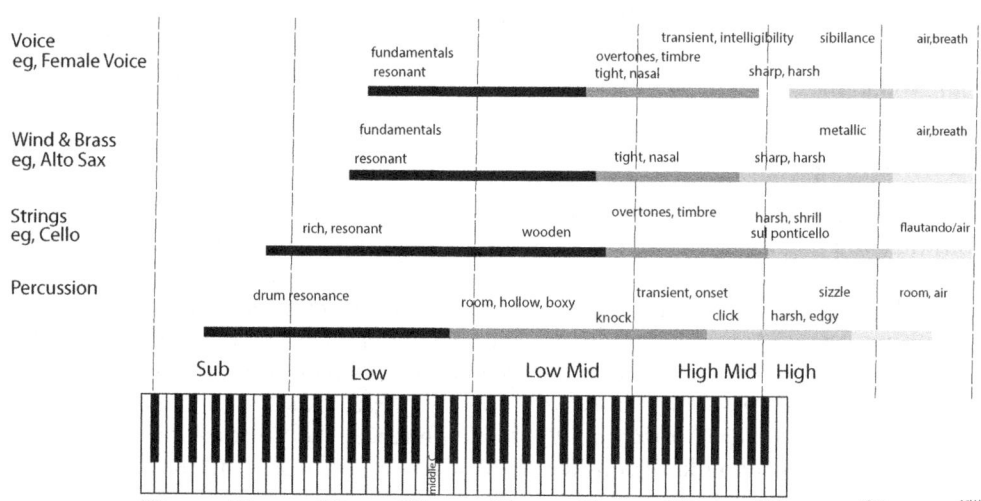

Figure 11.2 Characteristics of sound over frequency range

Within a single sound you might also consider this readjustment. If you take away 60 Hz from a kick for example, you might boost just above that. There is also phase shift at work with this technique (refer back to Part Two where we discussed how phase and filtering work). This concept is the basis of the Pultec style EQP1 low end filter where you can boost and attenuate the 'same' frequency.

Repositioning in space

Remembering the starting point of this part of the book is about decisions, and acknowledging that not everything can be up the front in the centre. Often when theres muddiness it is due to lack of attention to the relative position in the listening mix space. By moving sounds around in this space, it is quite astonishing how it can help individual sounds shape and character to emerge from the mud.

We have talked a lot about the timbre/frequency of sounds and trying to separate them through this lens, but we need to also remember that sound is always working on multiple dimensions.

Simple balances

First of all, consider the stereo field and panning. Secondly, the simple fader adjustments combined with a frequency change can be a good combination

Shape and dynamics

Considering the shape of sounds and considering how they may overlap or interrupt each other is something that might be important to consider when trying to 'unstick' sounds from each other.

Phase/space

Finally, you might consider working with a reverb or some kind of phase processor to move the sound in more subtle ways. For more information about this, check out 'I feel it all around me'.

Side chaining

Is side chain the answer? You notice that I have put this point at the end of a line of options that you might like to explore. I always consider side chaining to be the last resort rather than the first option. Side chaining everything is lazy mixing in my opinion. It also means that you not making any decisions or conceptualising the mix, and not being critical in your engagement with the sounds. The chances that your mix doesn't sound good or is rather flat are much more likely. Remembering also that when things feel 'stuck in the mud', one of the issues is that sounds are stuck together, side chaining by its actual process bonds two sounds together. It actually enhances the 'sticking' rather than unsticking them. When you have considered the above and the issue still persists, the two parts need to live together, then your only option is to side chain.

Chapter 12

Fighting for attention

The topic of fighting for attention tends to be more about upper frequencies. The attack transients and upper harmonics, and area that feels like an assault when there is sonic conflict occurring. We have covered lower frequency sounds that cover each over and give a sense of muddiness when not working together, so now we focus more on the upper parts. Sometimes this is an arrangement issue, other times it can be the result of the way the sounds were recorded. In both cases, there is quite a lot that we can do for fixing this in the mix.

What are we hearing when we are listening?

- A sense of messiness in certain parts could be harmonically, some kind of overtone dissonance, but often it is heard as some very subtle rhythmic messiness
- Often it is in the front part of the sound – there might be rhythmic messiness, or just a feeling that multiple sounds are trying to get our attention at exactly the same time and the brain is confused about where to go
- Sounds of similar spectral content playing at the same time, often with different rhythm/melody
- Often there will be a certain 'unpleasantness' or harshness, it might hurt your ears
- You might feel a ringing in your ears after listening or get a neck or behind your ear pain
- You might also be taken out of the flow of the song in the section where this issue arises

Concepts covered in this section

- Frequency
- Transient
- Envelopes
- Envelope shapers
- Muting
- Side chaining
- Multi-band processing

DOI: 10.4324/9781003289234-15

Frequency

Sometimes the harmonic arrangement has some issues. You might find that different instruments are playing in the same register. This can create a conflict where the ear is trying to prioritise one part over the other. One way we can combat this is to adjust the register of one of the parts (i.e. move one part or other up or down an octave)

If moving it up or down an octave is too dramatic, we can achieve a similar result by adjusting the frequency balance to focus the ear's attention on a different point in the sounds frequency spectra (harmonic makeup) rather than fighting the other sound at the fundamental.

Transient

Other times there might be instruments that are articulating different rhythms that seem to be getting in the way of each other.

After considering balances there might be some moments where two instruments have similar importance, i.e. a lead vocal and a guitar solo. There will always have to be a decision on which gets the highest priority. Not making decisions is a big factor of messy mixes.

Once we decide which has priority then we need to be able to move one slightly out of the front, but often beginning mixers will think that compression or fader adjustments is the way forward but after doing so the problem remains.

How do we move the transient of one sound out of the way?

There are multiple methods to move the transient.

Stereo field

Sometimes panning is all that is required!

Time and space continuum

> Often if we employ simple phase processes such as Haas where we apply 20ms approx delay to the sound creates a perception of moving away from the centre.

Diffusing the transient

> This is a similar technique to above, but we might use something a bit more dramatic such as a chorus effect in parallel that will create an ever so slight 'messiness' as the transient is doubled, pitch shifted and modulated this has the effect of diffusing the transient, and thus reduces the prioritisation that the ear gives to the transient.

Controlling or reducing the transient

> This method uses compression with a very fast attack and fast decay to clamp down on the transient. Be careful with this method it is very easy to do this poorly and end up creating an unnatural suffocated feeling to the sound signal. Make sure you have loaded in a very fast compressor for this job or use a transparent shaper.

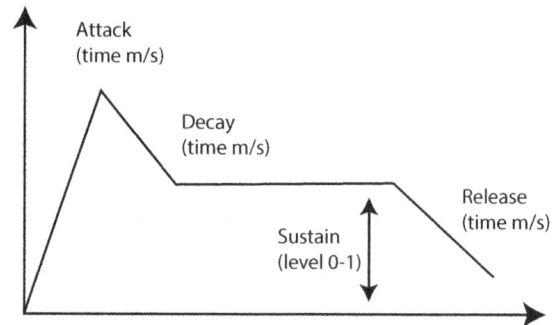

Figure 12.1 ADSR diagram

Envelopes and shapes

What are we talking about here?

Every sound has an envelope (shape). If we consider this from the perspective of sound design or synthesis, we might analyse the envelope or shape of a sound by analysing the ADSR (Figure 12.1).

Let's consider the visual representation of some sounds (Figure 12.2):

Some sounds have a clear on an off point and others don't. The sounds with strong onsets sounds will generally be more rhythmic it is easier for the human brain to clearly assign an onset and thus create a relationship with the next onset from the same or different instrument thus creating rhythmic patterns. It is primarily these onsets and offsets that we are concerned with in this section.

We might notice that a single sound seems to feel loose or out of step with others such as a long boomy kick drum. Or might notice that a group of sounds sounding together feels *messy*. Some examples might be: stacked kick drums, vocal harmonies.

What tools can I use to adjust envelopes?

There are a few techniques that we might employ to fix this. Firstly, we can use our envelope shaping tools

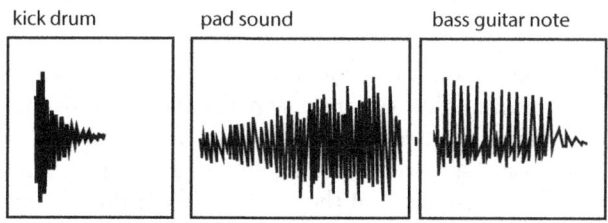

Figure 12.2 Visual representation of sounds

Compression

We might use a compressor to refine the contour. A fast compressor is key here. Take time to explore how the attack and release settings influence the envelope.

Transient shaper

We could also use a transient shaper, this I find particularly helpful on the front part of sounds.

Gate

The gate is a more extreme version of a compressor, it will decide what is allowed through and what is held back, combined with the attack, hold and release, it can be a very powerful transient shaping tool. If the gating effect is too drastic, remember that you can most often set a gate in its downward expanding mode that will let the gated sounds remain but at a much lower volume level.

Manual adjustments?

We might consider manual adjustments of clips/regions by performing fades. This is particularly useful for the occasional issue on an acoustic human performed track

Further technical solutions

Muting

The power of muting is one that many people forget. It seems so drastic! My feeling about muting is very simple: if when you mute the part you do not notice that the sound that was there is no longer there, then it was completely masked by other sounds and therefore if the listener can't hear it why is it there? You do need to be careful with this, if you ears are not sensitive enough there might be something that it is adding relative to other parts, and if you are working with a producer, then they may be able to detect that it is gone.

Often times I find that the stacking of layers in the low end is a producer trying to get big fat sounding kick or bass. Once I am confident of the sound that they are going for, if I can achieve it more easily with less layers, then I will go ahead and do it! It will help me later when I'm trying to get the maximum power (see this section coming up next)

Saturation

Some people say that the reason old records sounded so great was due to the unavoidable saturation inherent in analog systems, in particular the old school tube and tape days. Saturation is a magical process in many ways. It changes the tonal balance and relationship, it soothes out high end as an artefact of both tube and tape saturation is high frequency role off. This is perfect if the higher parts of the arrangement are fighting for attention.

Side chaining

Once we have tried all the above points and the conflict remains, we have to find a way to make the parts work together. Now we get into the more specialised processes. These are more advanced and are not the types of things that I would want to do often in a mix. When you are needing to do things like this, often you have realised that there are issues with the arrangement. If you can fix it in the arrangement, do it!

Multi-band compression or dynamic EQ

When we have only part of a sound or collection of sounds that are not working well together, we can use these more advanced tools to control parts of the sounds dynamic range over specific frequency bands.

Furthermore, we can use the **side-chain** keying process to further specify that the specific issue only be keyed out when the other sound is playing.

We can use either multi-band compression or dynamic eq to find the problem area and duck it out when it is getting in the way of another part.

Whether to use multi-band or dynamic? Relates to whether you have a specific resonance, whether you want it to be more focussed around a specific frequency or whether you want to control an entire band in a similar fashion. A dynamic eq allows you to work with filter shapes for more specific and targeted use, a multi-band compressor will control a wider band width in a more uniform way, so is good if there is a larger range that wants to be controlled.

De-essing

Single band ducking of problem frequencies is not just good for vocal sss. A one-time secret tool of mixing greats, a de-esser has been used to reduce the string noise from guitars and other unwanted artefacts that occur within the bandwidth of a de-esser. Remember what a de-esser is! It is just a narrow bandwidth compressor. It might help to consider it as one band from a multi-band compressor. The bandwidth is usually from about 1k to 20 kHz.

Chapter 13

More power!

This is another request that is often said to mixers. Clients want the mix to sound more powerful. Bigger, have more impact be more epic. Be louder!

What are we hearing when we are listening?
- The mix looks like it should be loud (i.e. meters maxed) but it is significantly quieter than other mixes even when you put limiting on
- Often other issues are going on as well, such as no depth, muddiness, lack of clarity or ability to distinguish the individual elements
- It might be harsh due to over compression over saturation in the hopes of making it louder
- The onsets might actually not be as impactful as other mixes
- The sub/low end might feel tight and not impactful

Concepts covered in this section
- Phase
- Arrangement
- Frequency masking
- Dynamics
- Saturation

Phase issues
The first thing to check if you feel like your mix lacks impact is whether there are any phase issues in your mix. Phase is somewhat of a mystery to most people when they start with sound engineering. People talk about it but few know what they are going on about! Now if you have done the preparation sections of this book, you will be well acquainted with phase. If you haven't, perhaps now is a good time to go back and review.

Phase or comb filtering is largely characterised by an unnatural tonal distribution, a weaker feeling in the mid-range, a sense of a hollowness and when stereo, some weird things happening to our perception of space in our mix.

So what is phase cancellation or how does it happen?

Phase cancellation occurs when two of the same signal playback out of phase (not time aligned) and it results in cancellation. This might be full cancellation or partial cancellation. Full cancellation only occurs when an identical signal is played exactly 180 degrees out of phase. Otherwise, the most common occurrence is partial cancellation. The result is a sound that is often described as `phasing' (it is also known as comb-filtering). Recall our study in Part Two of the book, if it is not clear to your what phase is, it is important that you go back and study the earlier section further.

Why does it occur in the natural world and in our music?

Most often the reason is that it occurs in recording is when two or more microphones have been used to record one sound source (for example stereo piano, or multi microphone drum recording). If the diaphragms of the microphone are not equidistant from the sound source phasing will occur.

You will remember also from Part Two and our discussion of equalisation that phase distortion can also be an artefact of the process of equalising. Phase can also happen elsewhere. We can create the phases sound on purpose for effect by using a phasor or chorus effects (see Part Two on phase processes). Using a phase correlation meter can help in diagnosing it. Once this has been detected it can be fixed through manual adjustment of the time between the audio files, or by using a phase alignment plugin.

Arrangement problems

(see also; not everything can be front/centre)

When we consider power issues, we sometimes realise that the source of the issue is in the arrangement of the music itself. Having too many instruments playing the same pitch notes can create a build-up in particular this occurs in the low-mids. But this might also happen between an electric guitar and a busy female vocal part – they will have a lot of energy in the 1–4k region if both are trying to get our attention at the same time, it can create issues.

We will feel problems in arrangement that have the potential to impact the perceived power of the mix when the build-up is in the low end, which means, sub, low and low mid frequencies. Or from about 20 Hz to 600 Hz.

Are there three kicks, sub bass, low synth pads, mid bass, male voice all happening at the same time? – that is going to have a great deal of information/sonic energy focussed in the low range. Without prioritising sounds in this range, the mix will look loud in the meters but might not be as perceptively loud in comparison with other music. This is due to a concept called masking.

Frequency masking

(see also; stuck in the mud)

The example given above where we have kicks, sub bass and maybe a low pad synth without any processing means that there will be a lot of information sitting in the low end. It is unlikely that every detail in the frequency spectra of each of these parts will get through to our ears. The likelihood is that one or two sounds (the loudest and the strongest onset) will take priority to our ears. Perhaps the low-mid melody line is getting lost in the mix. Starting mixers often come to me confused saying 'when I solo it, it sounds loud but when I play it in the mix it is barely audible'. Whenever this type of situation occurs, you can be sure it is frequency masking. We have already talked about frequency masking in detail (see stuck in the mud), so why are we talking about it now. If it goes undiagnosed in the low end, it can mean that there is a lot of low energy build-up in this area. Rather than resulting in a feeling of loudness, the opposite occurs because sounds are being covered over. If you have ever had a mix where it looks like it is loud, in the meters but compared to the commercial releases it does not sound as loud, low end frequency masking is likely playing a part.

Equalisation to reduce the frequency masking is required. We have already looked at this technique in detail in stuck in the mud section. Please refer back to this section.

Dynamic processing

Often, the first stop in the quest for loudness is to reach for the trusty compressor. An important thing to note about compression is to remember that it actually makes sounds quieter. We have turned down certain elements of a sound that are louder than our chosen threshold value. After these have been reduced, we end up with more head room (space between our loudest sound and the system maximum). So that means that we can turn it back up to the level it was before compression. Now because the dynamic range has been reduced, this same level appears louder because the signal is denser more compact. (For more information on compression and compressors, check back into the tools section.)

You will note that I talked about frequency first before beginning the discussion of dynamics. Compression is certainly an important part of the process, but it is not the only thing to consider. If you just go straight to compression and don't consider the impact that frequency issues (as discussed above) can have on the signal, the issues may remain.

Envelope

When we think of compression or dynamic processing, we generally think of amplitude or loudness but is important to remember also that these are time-based processes, so they change over time as our music is changing overtime. Once we understand that dynamics are time processes, that they work on (or with) the envelopes or the contour of the sounds then we realise that we should also be thinking about the envelope of the sound when we are working with dynamic processes. Setting and sculpting the envelope through attack and release settings on compressors as well as on gates, and even devoted envelope shapers can help create a harmony between the elements and proved a control or tightness that helps support a focused powerful mix.

(Also see; let's get this party started.)

Saturation

Whilst the compressor might be the first thing we consider when we think about controlling dynamics, there are also other processes which combine elements of compression in their process. We might also consider saturation. Tube saturation and tape saturation are the most common and both are useful when we want to help our sounds feel louder and more powerful. Saturation is great because it combines dynamics control with tonal shaping and works very well on low frequency sounds.

(Keep on reading because we get more in depth in the next section – thicken up the sauce.)

Chapter 14

Thicken up the sauce

Sounds individually might be lacking character, they seem thin and they need more 'body', to help them take up more space in the mix.

What are we hearing when we are listening?
- The mix might sound a bit 'polite', or boring, or not exciting
- There might be too much dynamic range
- Individual sounds might be being lost
- The mix might feel a bit flat and lacking depth or body
- You might want to make the mix perceptively louder

Concepts we will cover in this section
- Using saturation
- Using space

Using saturation to create thickness

It is worth reviewing the section that discusses the different types of saturation that are used when mixing. The assumption in this section is that you have a good working knowledge of the differences between tape and tube saturation and understand terms like 2nd- and 3rd-order harmonic distortion (THD)

Individual sounds

When you have an individual sound that might have been recorded poorly or just generally, you have a feeling that you want a sound to be bigger saturation is often a great help.

Depending on the type of sound, you might try out different methods of saturation. I often find that tube saturation tends to work better for low frequency sounds (like bass) where as I like tape saturation when I want to tighten up the mid-range of a sound. Certain sounds like snare's love some saturation to help focus and tighten the low resonance and help the control the dynamic range without feeling squashed.

Groups of sounds

What do we mean by groups of sounds? Well this would likely be the sounds that we have already decided are working together and we have routed or bussed them together. For example, it might be a drum bus or a vocal bus. It could even be the whole mix.

Saturation is often a really effective way to get that 'glue' feeling helping individual sounds be heard by the listener as a cohesive texture of sounds, and in doing so, it also can make them feel thicker, more impactful, have stronger foundation in the mix. In particular, tape saturation tends to work really well for groups of sounds.

Side effects

Remembering that for everything you do you often have some type of artefact, you can decide if these artefacts are helpful, or whether they detract from the sound. For example, when trying to help a sound cut through a mix, if I'm not listening to what part of it wants to come through I might mis-diagnose it as low mid when its actually transient and use tape saturation but not thinking about the fact that tape saturation tends to roll off upper frequencies.

Not all saturation is created equal!

A small note here to take your time and explore some good saturation options. Not all saturation options are equal or are designed to do the same thing. If you feel that you are getting too much signal degradation when you ae using your saturation tool, there is a good likelihood that it is modelled on a lo-fi analog unit. This is designed to do a different job and will not have the impact that we are going for in this section. It is always good to find some recommendations from your favourite mixers. Then take your time to demo these and explore (with your ears) how they work and how they affect the sound. Refer to Part Two and make sure you have your tools selected well before you find yourself in the middle of a mix.

Using space to create thickness

Another way that we might consider thickening up an individual sound is by changing our perception of the size of that sound. This relates to changing the perception of the body of the sound. In other words, considering the physical body of the instrument that may have made the sound (eg, the size of the drum).

When we use this method effectively, we are using 'time processes' (delay/reverb) to give us a sense of depth or dimension of a sound and help the brain to perceive the sound as 'taking up more space' than it did before. This is a very abstract concept, so if it's confusing don't panic. Often times the words make it feel more complicated than it is. My advice would be to try it out and use your ears to understand this concept.

Using the right tools is key

When we are exploring this concept, it is important that you have the correct tools for the job. The types of reverbs that work well for this process would be rooms and chambers. These have short quite clear resonant characteristics. If you are exploring this with delay, consider the way that the delay colours and potentially modulates the sounds and decide if you want this. Feel free to explore this effect with different tools sometimes it can be trial and error to get the right one.

Make it short and no tail

You can start with presets. We are going for very short spaces. Rooms and chambers. Often be careful not to select anything too bright. You might also manually adjust the pre-delay and the time to find the exact sweet spot.

Whatever you chose, you don't want late reflections, or feedback and multiple taps so be sure to take time to craft the effect manually. Often you will also need to apply filters (eq) in order that the processing doesn't change the spectral character of the sound.

Insert or parallel?

With both saturation and time/space processing there are instances when both insert or parallel would work. My advice would be to try it out but to be aware and listening for unsatisfying artefacts with both approaches. Following I will discuss a bit about how I think and listen when I am searching for thickness with both of these methods.

Insert for source bonding

What I mean here is that when an effect is inserted directly on a channel the resulting effect blends more fully with the sounds, so we don't hear the two things as separate. This is a good method if I want to change the entire sound. For example, take a thin snare to sound thick.

Remember with this method you will need to find the right balance of wet/dry.

Parallel for less artefact

Parallel to keep the focus and quality of the original intact and add another layer within the mix. This is important to use when you need to keep the exact quality of the original sound, but want to give it more thickness within the mix, in relation to other sounds. Great for vocals that need to be clear and distinct but sometimes needs some more power when the whole mix is working.

For any parallel processing, you are likely to go harder on the effect and blend it in more gradually. So for parallel time/space, make it 100% wet, for saturation feel free to dial in the drive harder. Then you can blend these into the mix.

The benefit of this approach is that you keep the integrity of the original sound but when this sound blends with the mix the parallel channel give it an extra dimension.

Same but different?

If you have read this section and feel like I'm splitting hairs, well, particularly with mixing, the devil is in the details. We need to pay attention to the small things, because these small choices can and do add up to have big impact on the mix overall.

There are no rules, but to listen and let your ears along with your skilled and learned judgement guide you. Or if you are lacking confidence or direction it is always helpful to take inspiration from those sounds you love and the people who made them.

Chapter 15

Let's get this party started

This section is very relevant if you are working with groove-based music. So that is anything that wants the listener to move. That means dance music of all forms, but not just electronic music, any ensemble that wants to get the listeners moving and grooving.

What are we hearing when we are listening?
- The groove doesn't feel cohesive
- The mix might be technically good, but a bit boring
- The track doesn't make you want to move
- The mix might feel a bit messy or lose

Concepts we will cover in this section
- Getting into the groove
- Envelopes and envelope shaping
- Compression and envelope
- Gating and downward expanding
- Side chaining
- Psychoacoustics

Getting into the groove

I called this 'let's get the party started' because it is about ensuring that the elements of the mix are working together not against each other. We want to ensure that they are working together in the correct way to communicate the intention of the groove. To further articulate this point, I like to use an example comparing techno to funk. Both are dance music that want the listener to respond through dance, but the type of dance they encourage is entirely different. If we don't accurately assess the type of groove, we can work against the goals of the production and result in a bad mix. So, take time to assess this before diving into processing here.

Envelopes and envelop shaping

Now that we are thinking of the groove, we might consider the shape and the movement of individual sounds. For techno, I would be considering that my kick drum is powerful but not long, the tail of the envelope of each kick needs to be finished before the start of the next one. There are several ways that we might consider shaping this. Firstly, there is a specific tool for shaping the envelope called a transient designer. This tool can reduce/increase the intensity of the transient and it can alter the release characteristics of the sound.

Compression as a method for envelope shaping

When we think of compression or dynamic processing in general, we often just think of amplitude or loudness, but is important to remember also that dynamics processes are time based processes. Once we understand that dynamics are time processes, that they work on (or with) the envelopes or the contour of the sounds then we realise that we should also be thinking about the envelope of the sound when we are working with dynamic processes. Working with the attack and release of compressors can drastically change the perceived envelope of a sound.

By manipulating the attack and release setting of a compressor on a rhythmic, repetitive sound we can start to hear how different settings bring out different feelings or grooves in the repetition.

Exercise: Try this yourself with a kick drum in different genres and see if you can hear how these manipulations can impact the envelop which changes the groove.

Gating and downward expanding

Gates are a very important tool for tightening the envelopes of sounds. It is often used to keep out background noises or bleed of recorded sounds. Some examples might be room noise of a vocal recording or other drum sounds that have bled into a kick or snare track.

In certain types of music, we might want to create a more artificial sound so we can use a gate to cut out any other sound from the kick microphone, but sometimes that feels a bit intense and makes the drum groove off kilter, especially in more smoother types of patterns. In this case, we might use downward expanding instead. Downward expanding means to increase the dynamic range downward. We want to lower the noise floor. We set it up in exactly the same way as a gate, but instead of having silence after the gate has passed the threshold and the gate has closed, we still hear the sound but it sounds much quieter. We determine how quiet by setting the range. This way the bleed sound is still there but the distance between this and our 'direct' sound is greater. This gives us a more focused direct sound but allows it to keep the natural feel. When we move to compression, we start with more space between the two sounds, which thereby keeps the noise floor low and allows the compression to work on the sound rather than the unwanted background noise as well.

Connecting elements to each other through side chaining and envelop following

So far, we have been considering altering the shape of single sounds, but this section is also about considering the relationships **between sounds** as well. We are not only considering the kick by itself, but also we should consider the kick to other elements of the track. For example, what is the relationship of the kick to the high hat? Or the snare?

And what about the different high-hat patterns and how they might interact with a more ambient sound in the mix?

Side-chain compression

Most people in 2024 have heard of side-chain compression, it was made famous by French band Daft Punk, using a kick drum to duck a pad sound. In this method, we would send the kick drum to a compressor (that has a side chain or key input) on a pad and using the envelop of the kick to cause the compressor to engage and disengage.

Side-chain gating

This process is similar to side-chain compression but rather than ducking the sound we can make more dramatic changes and actually turn off, or close the gate on parts of the sound. We can also invert the gate and create inverse effects for more complexity and fun. Gating works better when the modulating signal has a shorter envelop.

Exercise: Try putting the high-hat signal into a gate that is on a more ambient sound and see how it makes this ambient sound move with the high-hat rhythm. Also experiment with inverting the signal.

With both of these techniques, we can go wild with them in production, but in the context of a mix, we could use these in a more subtle way, that doesn't perceptively change the character of the sounds, rather it creates space and movement between the elements that creates at the same time more clarity of the parts as well as enhancing the connection between them.

Psychoacoustic phenomenon of 'streaming'

When we use a side chain, the process uses the signal of one sound to control the settings of a processor working on another sound. This starts to fuse characteristics of these sounds. We can understand it in some way through considering the psychoacoustic concept of streaming. The brain is constantly performing perceptual shifts and organising sounds into different streams. This shifting may be one of the reasons we find enjoyment from listening to music, constantly experiencing different relationships as we shift and create different streams and relationships between the sounds in the music we listen to. Mixers (and producers and composers) exploit this concept by the way they approach arrangement, orchestration and sound processing. As a mixer 'keying in' the envelop of one sound to control parameters of processing on another sound is a great way to create a bonding or relationship between the sounds that can encourage a connection or a streaming experience for the listeners.

Why would we want to encourage streaming of rhythmic elements?

In dance music, much of the power of the groove is through the way that the elements play off each other, enhancing the relationships between the sounds, at the same time as we try to create a clarity of perception of the individual sounds is much of the magic of these genres. The way that these sounds play off each other creates the particular groove. The envelop, attack and release, is at least in part responsible for the different types of grooves.

Chapter 16

Low-end theory

This section is devoted to talking about issues involving the lower end of the frequency spectrum. Balancing, tuning and making the low end strong and powerful is one of the most challenging aspects for aspiring mixers. In the following section, we go through challenges and different techniques and approaches that can help solve the low-end issues. In many ways, this section is a retracing of many of the steps we have already taken in previous sections. We go over it again because of the importance and challenge of the low end. So let's get into the low-end theory and make those subs rumble, kicks hit and bass line melodies shine.

What are we hearing when we are listening?
- The low end doesn't feel focused
- The mix looks like it is loud on the meters but compared with other tracks it is not perceived as loud
- One element or more of the low end might not be cutting through – i.e. mainly hear the sub-bass and not the kick, or can't hear the sub, etc.
- Perhaps the mix is ONLY kick and bass and the other elements seems to be lost when the kick and bass are working

Concepts we will cover in this section
- Challenges that block our efforts
- Relationship between the parts
- Don't get stuck in the mud
- Side chaining
- Dynamic processing
- Envelope
- Saturation

Challenges that block our efforts

Before we get started on this section, it is important to check in about some of the challenges that can impact your ability to make sound decisions on you low end as well as remembering ways that you can work around these issues.

DOI: 10.4324/9781003289234-19

The big one is that the listening environment is not optimised for mixing, which basically means you are working in a standard room of a house or apartment that is not treated and therefore has standing waves which either enhance certain frequencies or remove certain frequencies in the room.

The second one might be that you are not using professional monitoring and the sound of the speakers is colouring your perception. In particular, if you have small monitors you might be thinking, how low does you speaker go?

Solution

These issues have haunted every mixer at some stage of their development. The ultimate solution will be to invest in your monitors and also treating the room. But if you are not at that stage financially or practically yet, it doesn't mean that all is lost. It is not like you can't mix until you spend 10,000euros on an optimal room set up, it is just a lot easier to mix once you have optimised your listening environment. In the meantime, the main solutions I offer are as follows.

Use references

Songs that you know well that you can play in this room and judge the tonal response of the low end, and this helps you to make relative decisions about the mix that you are currently doing. It needs to be a song you know well.

Learn your room

You can perform some tests of the room and find out the problem areas and then be careful with these when you mix.

Develop your ear sensitivity

Listen to references within this room, and learn the room through your ears.

Room aside, you might not have developed the critical skill required to accurately hear the sounds, or to diagnose the issues, in particular in the low end. The solution here is to practice (go back to my listening section as often and as long as you need to).

Remember that in Part Two of this book there is a detailed section both topics of listening environment as well as listening skills, so be sure to go back to these and dig in if this is something that you need to consider.

Ok so now you are all set, we can begin working on the low end.

Relationship between the parts

As with everything in mixing we are mainly listening and evaluating the relationship between various parts in the arrangement. A quality arrangement will have thought about these elements already, but often it can be ambiguous.

What do we mean by the relationship? Well that are the roles of the instruments. For example, is the kick in the mix for a low-end thud, or does it also have a purpose to have a strong sharp transient? Does the kick want to 'knock' or does the sub-bass want to push in that region to bring out the melody of the bass line more clearly? Which element needs to have the most space in the sub-region?

When we consider these issues, we may take a variety of approaches. Let's look at some of these now.

Don't get stuck in the mud

That's right, consider whether there is frequency masking. Consider your arrangement. Can things be adjusted in the arrangement? Can movements between register occur? If not, then we consider eq to prioritise sounds in different frequencies.

A common approach is that we will analyse the kick and the bass parts, looking for where the fundamental sits in each. The kick might be 45 Hz, with an overtone around 90 Hz. The bass might have its fundamental around 60 Hz. If this is the case that is quite good, we will have more problems if both parts have fundamentals on the same frequency, (if the bass also has fundamental frequency around 45 Hz), we likely need to make some decisions and perform some cuts.

We need to make some decisions in this case. Does the song want the kick to hit in this sub-range, or the bass? Which takes priority? If we decide that the kick should have the space here, we would remove some of that frequency from the bass, we could then in order to give some low back to the bass perform a little boost on the other side of this range, for example at 60/80 Hz so that the bass still has that sub-frequency. We would then remove some of that frequency from the kick to make room for the bass. This give and take allows us to carve out some space in the low end for both of these low-end sounds.

A few considerations to be aware of. Phase in the EQ settings creating unpleasant results. With this one, just trust your instincts if it feels like it does something not nice, trust that! Use another EQ, slightly adjust the position of the notch or be gentler with your EQ settings.

Letting AI do the work?

We are in an amazing time with a large array of AI tools designed to make our jobs easier! These tools can help beginners consider their options, and it can potentially speed up workflow for any mixer. In Part Two, we discussed AI tools, and again in Part Four, I go into more detail about AI.

In the context of the discussion here about frequency masking and trying to clear away muddiness, there are many AI tools that might help you. Just remember that they are suggestions and ultimately you as the mixer need to employ your own critical judgement as to whether the suggestion works for your approach.

Side chaining to clear away low end

Another thing we should talk about here is side-chain processing where we might key the kick into the bass. This is a very popular approach in contemporary mixing. This method we take the signal from the kick and key it into the side-chain input of a compressor that is on the bass. This means that every time the kick fires it turns on the compressor on the bass. This has the effect of ducking the bass when the kick plays. Seems like an easy solution, right?!

Whilst it seems like a simple fast solution it should be approached like everything with consideration. In my mixing, it is generally my last choice not my first choice. I see too many students who side chain automatically without properly diagnosing the issues in the first place. Automatically side chaining can create other issues. It can create a diffuse muddiness in the low end, but more-so I find that it takes away the detail that I am after. I don't want to remove the bass entirely when the kick plays, I just want to carve out some space so that they don't fight for the same space. If you have bass and kick playing at the same time, they should probably be heard, the producer would have left the bass out if it didn't want it there.

I recommend analysing the low end of both parts first, as above with the discussion of equalisation. Find out where each part has its fundamental and first overtones. Then consider whether you might prioritise an area for each part. If both kick and bass have fundamental tones in the same place, then you need to make a choice here. Some eq shaping might be a first step, then you realise that you need more carving, by all means apply a side chain, but perhaps rather than a normal compressor, you might reach for a multi-band compressor or a dynamic eq where you can pinpoint the area of the bass that you want to reduce when the kick fires. This way you leave untouched the areas of the bass that should stay in place when the kick plays.

Identifying phase issues

Producers often stack multiple kicks, sub-bass and bass elements together in their arrangement. They are wanting to create a powerful and thick low end, but sometimes these layers can actually be working against each other. If two samples are out of phase with each other, i.e. one part is pushing when the other is retracting, we will have a phase cancellation which results in loss of sonic power.

An easy way to identify phase issues is to play one sample in solo and then add the second one, if it sounds quieter when both are playing, or somehow weaker then you know there is a phase issue. The next step would be to zoom in on the samples and look to see the way the waveforms are moving. They won't be perfectly aligned because they are complex and different waveforms, but you can make adjustments to make them more aligned. Play the sounds as you do this to listen to the adjustments. Another way to identify phase issues is to use phase correlation meters, Panning the samples left and right then reading the correlation meter as it considers the relationship between the left and right channels will tell you if it is phase aligned or not.

(refer to the discussion on phase in Part Two of the book if phase still confuses you)

Dynamics and envelope

When we are thinking of the low end, we are considering the vertical axis but we must also remember to consider the x or the time axis. Paying attention to the vertical axis first we utilise dynamics to ensure a balanced amplitude across the part. Be careful not to overly flatten it. The low end still needs a sense of movement. If you drive a compressor too hard, you can have the opposite effect that you are going for. If there are resonances, you are trying to control manual adjustments and things like dynamic eq would be your friend. We often work with compressors in our bass elements to solidify and support the sense of shape of each of the sounds. The transient onset and the resonance and shape of the tone must be carefully shaped between the low-end elements.

Saturation

Saturation makes a sound feel more dense or thick, particularly in the low-mids. An artefact of tape saturation (different speeds will have different results here) is a push in the low-mids, which is particularly pronounced when there is low and sub-information that is passing through. If we consider that saturation introduces overtones (even or odd) from a fundamental, it makes sense that if we have sub-frequencies present, then the low-mids are pushed – sound. For example, a fundamental at 100 has overtones at 200, 300, 400, 500, 600, 700, and 800. If the type of saturation is predominately 2nd-order (tube), we will get even harmonics/overtones occurring at 200, 400, 600, and 800. We can see that this is predominately overtones that reinforce the fundamental.

Let's take a specific example when working with low end. Sub-bass is notorious for being lost when played through smaller speakers, and even when monitoring through large speaker sub-bass can struggle to find its place in a mix that has lots of other low-frequency components. We can use saturation to pronounce the overtones the harmonics. This means that we create new tones that were not there before through saturation, these overtones, or even harmonics reinforce the sub-note, in a higher register. In this mid-range register, the human ear is more easily able to pick this up and that fun thing about it is that even though the ear might be registering 800 Hz first, generally speaking the 800 Hz actually just helps the listener get access to hearing the sub-bass. Psychoacoustics are at work here.

The second reason why saturation is such a helpful tool is that through the saturation process there is a dynamic range reduction that takes place. Acting as a compressor saturation can tighten up our low-end parts whether it is a kick or a bass and help to bring it under control and find its place in the mix without taking over everything.

Chapter 17

I feel it all around me

As soon as we add reverb and delays the music feels more alive, more engaging and more emotionally connected, but if we select the wrong types, or if we don't consider the relationships between the elements we can very quickly lose our way in the vastness of space and time! In this section, we discuss the element of space from multiple perspectives as well as looking at some best practice tips to help you create evocative spaces that work with your parts rather than taking them over.

What are we hearing when we are listening?

- The production, or elements might be lifeless, flat, like cardboard or just lacking sparkle and excitement. The snare might have no life, the vocals feel flat
- The mix might lack depth and feel like it is all existing on one plane
- The spaces used might seem to fight against each other or offer conflicting narratives
- The spaces might not match with other aesthetic markers of the production.

Concepts covered in this section

- Space and aesthetic
- Space and emotion
- Mix depth and width
- Creating a sense of immersion
- Leaving spatial processing until last
- Tuning spaces

Aligning spaces with the production aesthetic

This might seem like an obvious point, but it is something that I find myself discussing and recommending to student mixers. It is easy to be impressed or overwhelmed by the round of reverb on a dry instrumental or vocal track, but it very important to try to remain objective in the analysis of whether the sound and mood fits with the overall production aesthetic and emotional feeling of the music. If your song is intimate and confessional, a long reverb that washes out the vocal will work against that feeling. If you have a piece of music that is designed to feel as though you are listening to a live performance, then it is important to build spaces that match that experience.

DOI: 10.4324/9781003289234-20

Relationship between space and emotion

Space links to emotional states because when we listen we imagine ourselves in the spaces that we hear. It is a fascinating concept to explore. Furthermore if particular spaces or sounds are very closely aligned with previous experiences we have had it can create a strong referential relationship and can elicit even stronger associations. The concept is called source bonding and is another element of psychoacoustics that relates to the way we process sonic experiences. Knowing this, as mixers we want to be very careful to select the spaces that not only support but also enhance the emotional context of the music.

Perceptions of mix depth and width

Remembering that spatial processes are actually time processes, In reality, a sound that occurs further away hits our ears at a later point. We can use space in a way that is not perceived as an environment by the listener, rather it can create perceptual shifts in positioning to make things sound further away. For example, if I want a sound to float in the background of the mix, I would work with the parallel channel in pre-fader and balance the wet and dry signal to establish the relationship between the foreground sounds to get this sound to feel like it is coming from far away.

We can also use this for the subtle shifts occurring in the midrange. Subtly different rooms can move around the sounds within our virtual mix environment. Experiment with this yourself. Remember to keep an awareness of whether the room starts to be perceived as its own element (i.e. heard as a room) this would detract from the goal and could cause aesthetic confusion. This technique can be tricky but it is a powerful tool to harness.

Tips for working with space processing

Leave the space processing until you have cleaned and optimised the mix

This can sometimes be hard to do, but I find that as soon as I put a reverb on it starts to relax my attention to details. Things get lost in the wash, or the reverb convinces me it sounds better than it does. So, I started to focus on fixing and cleaning first before I applied reverbs and delays and it enhanced the quality my mixes.

Select the reverb that matches the style

For example, if it is an icy cold Scandinavian pop song I might use a contemporary digital reverb, if I'm going for a raspy vintage ballad maybe a spring reverb or a tape echo might yield interesting results.

We can use spaces to change the perception of the body or size of individual sounds

If the snare sounds thin, try using a small room that creates a thicker resonant cavity that can trick the brain into hearing it as snare body resonance. This technique can work well when inserted. It can work on many instruments, but be careful with the balance in case you create confusion with the transient information of the sound. You might need to perform additional equalisation after the reverb effect to retain the tonal balance.

Parallel rather than insert for cohesion and clarity

Rather than using multiple reverbs on inserts create parallel tracks and you can create sense of cohesion through creating a bus and sending multiple parts to the one space. This is great for multiple parts that want to be heard as one texture.

Have a plan for the spaces to ensure they are all working together

Though some engineers haphazardly use endless reverbs and delays most often it is helpful to be considered in how you build your spaces. Consider how the different spaces you are using not only work on the sounds that they are applied to but also how these spaces interact or work off each other.

Tuning the spaces

When we consider tuning the spaces, we can consider the frequency spectrum and cut out certain areas of build up or guide the perception of depth through careful sculpting.

We can also tune or shape the reverbs by calculating the decay and pre-delay settings to time relative to the bpm of the music.

Sometimes we might explore more experimental effects such as sidechaining into compressors or dynamic eq's for help with clarity of to create space and movement.

Chapter 18

They only listen to the vocals anyway

For most music that includes vocals, the reality is that the listener will be naturally drawn to the voice. Of course, if the listener is a musician with advanced ear sensitivity and training, they will be able to pay attention to other parts, but if the listener is an average person they will most likely be receiving the song from the vocal part first and foremost. This is because of the way that our hearing has evolved, we listen to voices to understand each other, to communicate; we are a verbal culture that prioritises the voice.

If you are mixing for experimental music, perhaps classical, jazz or ambient, maybe even dance floor music like techno, then perhaps this section is not one that applies to you. For everyone else, now we will combine all the elements that we have been talking about in the previous sections and consider the vocal. The vocal is so important and also the listeners knowledge and expectation of voice is more detailed. We all hear voices regularly and so it is incumbent upon the mixer to ensure we get it right, if we don't the listener will know something is 'off' about the track and the vocals.

The next section goes into detail of all the elements that we need to pay attention to in order to get a quality vocal production. We first focus on one vocal, the lead vocal and then we also talk about backing vocals considering the extra elements that we might think about relating to groups of voices, as well as the relationship between the main voice and the supporting voices.

The voice is a particularly challenging element for modern mixing for several reasons

In much contemporary music, it is often the only element that has been recorded, all other elements are either sample based or synth presets carefully designed and optimised by pack producers. This means that the vocal will need a lot of work to go from raw voice recording to fully processed sound that is equal to those pre processed sounds you use in your DAW.

Added to this is the change in recording approach. Back in the day if you wanted to record your voice then you had to go to an expensive studio that featured a specially designed room and hosted multimillion dollar gear. Now we can all record anywhere, such as bedroom recording. It unfortunately often means some unwanted sounds and tones can find their way into the mix.

Now that microphones and recording equipment in general is cheaper it thankfully means that so many more of us can make our own music. Whilst this is great, when we are on the path to create high quality commercial sounding music it can reveal limitations. All mics are not made equal and this can lead to issues with the quality of the recording of particularly voices.

Another point is that because we can all make music, we are all doing it. But again as we make music in our bedrooms and teach ourselves how to record, there can be bad habits, like not setting levels correctly as well as lack of sensitivity to detecting issues as they are being recorded. It is when we get to the mix, that these problems really start to come into focus.

Final point relates to the production phase. Depending on your aesthetic and genre the approach to processing the vocal can be mild or extreme. The productions that fall on the extreme end can often have lots of tuning, sharp EQ shaping and loads and loads of compression. So when we get into the mix, the vocals might have acquired some unwanted artefacts in addition to those things that the producer was going for.

Ok so with all this said, we can now get a bit more understanding of the myriad challenges that we might face when working with vocals, let's get into the detail!

Lead vocal

What are we hearing when we are listening?

- Are there any unnatural artefacts, clicks pops
- What shape is the sibilance in?
- Does the tone of the vocal feel strong and balanced, does the emotional delivery match with the sonics?
- Is the performance balanced?
- Do we hear all the words or are some getting lost?
- How is the timing? And tuning?
- Does the vocal sit correctly in the mix?
- What is the sense of space that the vocal is in? Does it work?
- Does the vocal sound professional, polished and finished?

Concepts covered in this section

- Cleaning vocals – including resonances, artefacts
- Dynamics methods
- Automation
- Use of space

Clean it up

When I ask people what do you do when mixing voice, the very common answer is people saying 'clean it up'. So let's get into what exactly that means. If we focus on frequency, we are considering the types of frequencies that might sound problematic.

Frequency

Clear away unwanted frequencies. Remember only do it if it is needed. You might need to clear away low frequency rumble, Within the actual vocal itself it might be too bright or edgy (upper mids), nasal (5 Hz–700 Hz), it might be boomy (1 Hz–300 Hz). For each of these elements, we need to be able to detect what the issue is, to be able to do this we need advanced ear skill. So if you are struggling to detect the frequency area that needs attention in the vocal, then it is time to go back to you critical ear training.

Working with unwanted resonances

Bedroom recording room mode

If the vocal was recorded at home in a bedroom or not specialised recording environment, there is a likelihood that room modes may have found their way into the vocal recording. This is often heard as an excess of low frequency which can be perceived as the 'vocal is boomy'. What is often going on is the voice excited a mode in the room where the recording was taking place and created a resonance at this point. Often the dimension of domestic environments means that there is a low frequency mode around 200 Hz. If you are recording many layers in this room, then this mode will be enhanced with every layer.

If you can notice it when you are recording and try to move around the room to find a place in the room where this effect is reduced, this is always great. But when it is already recorded, there is nothing to do but to cut it out. You might find yourself cutting it quite a lot!

Proximity Effect boomy-ness:

Another area that helps to look out for is the proximity effect. This is particularly enhanced in certain types of microphones, but basically if you are very close to the diaphragm it can have the effect of boosting the low frequencies. This can be the desired effect, but if we are stacking lots of layers together it can create build up. It is helpful to try to detect this issue when recording, have an awareness and a plan for different approach to mic position for lead and other supporting parts. Once in the mix, we can diagnose it and then will need to apply cuts as necessary.

Continual persistent resonances

A continual resonance that is present throughout the performance. This is generally created due to issues with the recording, such as a problem with the microphone or perhaps the room having strange sounds or creating unwelcome sounds as the voice reacts with the room. This would include the above special mentions of room modes and proximity effect, but it might also include other un expected artefacts.

For these issues, we would need to perform a cut, sometimes quite a strong cut. Remembering from Part Two when we discussed phase and EQ the more extreme the q the more extreme the phase shift issues, so it might be an option to explore using a linear phase EQ for this job.

Intermittent resonance

There might be resonances on some parts of the lead but not all. This might be due to the singer changing their position ever so slightly in relation to the microphone, depending on the microphone used this can create different tonal characteristics, and sometimes resonance. If we use multiple takes where the position relative to the microphone has changed, it can create inconsistencies on the vocal track. So we will need to remove these issues, but if they are not occurring all the time an eq that completely removes a certain portion of the sonic spectra is not the best option. We need something that can remove the issue only when it occurs – yes that is right it is time for using dynamic EQ (to review this tool remember you can go back to, Part Two).

Advanced tools for 'cleaning up' a sound

Dynamic EQ combines amplitude and frequency and only removes specific frequency when the amplitude hits a certain threshold. This means we set it so that it only reduces whilst the resonance is apparent and when not it should not be working at all.

AI tools for mixing have blasted onto the scene in recent years. Without needing to go into the exact details of how these work, we can gain an understanding that an AI resonance tool is able to be more thorough and adaptive in its analysis and treatment of audio spectra. This means that the resonance reduction is generally found automatically and/or is reduced actively at the rate that is required. In short, it takes out much of the work of having to set up the process manually. There are a range of tools out there, one very popular example is Soothe. This tool can be used to detect and reduce dynamically the resonances across the entire bandwidth.

Clicks, pops, mouth sounds

We have been focused on resonances so far, but the issue might not be resonance but actually relating to transient part of the sound, the front part. This can be a hard one to hear. This is also a phenomenon that would not have been an issue back in the days of analog recording. One of the artefacts of contemporary high-definition digital recording that is up to 96k and 32 bit is the level of detail that we can now get. In analog studio running tube microphones and mixed to tape, there is a significant smoothing that occurs on the high end, as well as the front part of the sound(transient).

So whilst we now have all the detail the question we often ask is 'do we actually want it?' Sometimes yes, sometimes no.

In this instance, let's talk about when we don't want it. The clicks and pops of a vocalist moving their mouth can actually be quite off putting. Even if it is not off putting it might be taking up energy and making our audio processing to be working too hard or in the wrong way.

When the vocal feels a bit sharp or not pleasant to listen to and you have already shaped the frequency spectrum, perhaps it might be the sharpness of the transient peaks. For these times, my go to is actually AI reduction tools. iZotope does a very good job with their de click and mouth de click. Essentially these are an adaptive gate.

Dynamics

Once we have cleared and cleaned up the vocal then next big thing that people consider with vocals is to try to focus or control the performance. Vocals are one of the most dynamic instruments that we have and it is no wonder that compression is often featuring heavily with vocal mixing. It can be a tricky instrument to control because the full range of the voice can be very quiet sound of a whisper, to the loud belting sound of an explosive soaring long note. Managing all of these aspects to fit somehow in a mix with a perception of a changing character but maintaining a balanced listening experience is one of the great arts of mixing and vocal production.

Selecting a compressor

One of the first things I consider with a lead vocal is what type of compressor(s) might be useful for this vocal. I might think of the style of music, but often times I'm more interested in the quality of the voice and the quality of the recorded sound. If I feel like it is very 'spiky', I might select a compressor that will help smooth things out. If it feels a bit dull, I might go the opposite way to encourage movement. In general, I might also be trying to help the voice feel more focussed, tight in sound.

This is often mis-interpreted that we want to control the peaks of the performance for example when certain notes are too loud, and whilst this to a certain degree can be part of the job, it is actually not really what we mean when we want to control the dynamics of the sound. It mainly relates to the sound envelope, the component parts of the sound's characteristics and focus the overall shape of the sound. Rather than the discrepancies of the performance (there is another approach for this which comes next). This distinction is subtle at first, but it is a very important one. Without paying attention to this you might find yourself over compressing your sound, and not paying attention that the compressor is not helping the sound feel fuller and more focused but is actually making it feel claustrophobic and flat.

Clip gain automation

As mentioned above compression in mixing is not really used to balance the extreme fluctuations in a performance. Often mix engineers are manually drawing in clip gain automations or cutting and adjusting clip levels. This is done BEFORE the processing, so it doesn't work if you automate the volume fader on your DAW. A common approach is to work with the compressor open and dialled in and help to determine which elements need adjustment (particularly the overblown notes) by looking at the meter. Clip gain automation is not just about turning things down, often there are words in the performance that fall below the track, and the words get lost. These need to come up. So it is important to note here that clip gain automation should occur in the mix and not with the vocal soloed.

Two-stage compression?

In more recent years, as technology advances, as the trends towards ever perfect balanced and optimised sound clients are expecting music we have seen the emergence of an approach to vocal mixing that uses two compressors in the vocal chain. Generally, the first compressor is faster and controls the momentary elements of the vocal line. The second compressor is often a slower compressor that works to smooth out the vocal overall and create a sense of glue or continuity over the vocal phrases. My go to compressors for this would be something like a FET (1176) as a fast compressor and finish it off with an optical (LA2A). Whilst this is not something that you need to do, you are working in pop or hyper-produced genres and trying to get a similar sound, then it might be worth paying attention. Multiple stages of compression are likely occurring on the vocal lead.

Parallel processing compression

We know parallel processing for its most common use case with reverbs or delays. To recap on parallel processing – we use a send or a bus to send some of the signal off to another (parallel/auxiliary) channel which has the processing applied to it. Then we can blend it with the dry signal. When working with parallel compression, a common approach is to heavily process the parallel channel with the goal of bringing up the lower elements and reducing the louder elements, then when this is blended back into the the combination of the two combines the characteristics of both in with the original raw signal.

It has the effect when blended that the louder parts of the sound are maintained and the quieter sounds are turned up. We hear both the loud dynamic part of the raw signal and the quieter parts of the compressed signal that have been turned up significantly through the process of over compression. The over compression has pushed down the louder parts so much that they are hidden when blended in with the raw dynamic track.

It is kind of an upward compression concept if you will. We leave the peaks where they are, but reduce the overall dynamic range by turning up the quiet bits.

There can be some tricks with parallel compression, and it is important to take time to sculpt the parallel channel so that you get the effect of the low bits coming up, we don't want to just blend in an over compressed signal that doesn't effectively produce the desired result. If you want to dive deeper into this concept, look up the legends who exploit it with great creativity such as Andrew Scheps.

Add extra sparkle

At this point, you should have a very clear and well-rounded vocal, now the fun stuff begins, we can start to make the vocal sound even better.

Character equalisation

You might be thinking hang on we have already equalised the vocal. Yes we have equalised it to clear and clean up the part. You might also have done some tone shaping of the lead vocal at that point. Now when we are putting on the finishing touches to the vocal, I might consider whether using a more exciting equaliser for some broad tonal shaping could benefit the vocal. Choosing the right equaliser is the fun part. You might be choosing it based on the style that you are working on, or maybe you have specific EQ's that you love that fit your personal mix aesthetic. Many mixers famously use the Neve 1073 on vocal chains. Perhaps it is something you might try out. What character EQ's did you put into your mixing tool kit?

Saturation and analog warmth

You might just be after the gentle warmth and character that a certain piece of analog gear (or emulation) gives to a sound. Many mix engineers might run a sound through some gear without using it for its intended purposed. For example, LA2A optical compressor but not compressing the sound at all. This offers gentle saturation and colouration that is given by the circuit which contains tubes in the output stage.

After considering all of these options your lead vocal signal chain might look something like this (Figure 18.1).

Make the vocal larger than life

Phase and space(s)

Remembering that mixing has a lot to do with perception and little to do with reality we can work with spaces that can change the way we perceive the vocal. We might use ultra-fast delays, such as the Haas effect to make the vocal feel a bit thicker and have more body. We could use a slap-delay to help the vocal sit out on top of a busy mix, which is popular in indie and rock genres. We might use different larger spaces to create a sense of whimsy, or we might use a cathedral to help the vocal feel angelic. There are many ways that we can work spaces to lift the vocal to the next level both sonically as well as emotionally. It is very important when working with the vocal in this way that we check in again on the aesthetic identity of the track, and in particular of the singing style and ensure that the approach you go for suits this.

Spaces most often in parallel

Even though you can duplicate reverb on individual tracks, it is still more common to work with spaces in parallel. We should consider the way that working with spatial processing in parallel can create different sense of depth perception than using inserts. Of

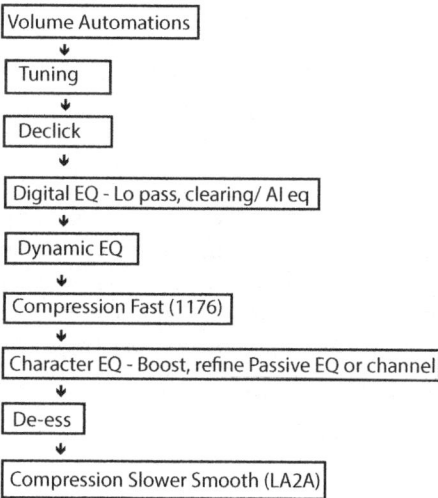

Figure 18.1 Example of a lead vocal chain

course there are exceptions but as a thorough mix engineer make your choices through consideration, connecting with the intention and most importantly listening.

How many?

It is a good idea to review the section on space in Part Two of this book where we go into detail on this. When crafting a lead vocal, it is common to have three to five parallel reverbs/delays working in tandem with the dry lead vocal. This is often a very short space, plate or Haas type effects that is not heard as space and rather adds some dimension to the vocal. There is often a room to make the vocal sound more natural. Often there might be a larger ambience, that may be automated for specific phrases or sections. There might be two different types of delays that are automated for sections and transitions. But of course this is just a guide there are no hard and fast rules. You can develop a mix workflow that tends to work for you, or you can make decisions in every mix based on the vocal and the aesthetic.

Automating or throwing to delays or more complex effects

I mentioned above about automating sends to parallel processing, so I think it might need some more explanation. This process would send the vocal to the parallel channel only at specific times. I can be used for sections to give a different feel or energy, it can also be used to highlight particular words. The benefit of the automation is that it also keeps the vocal part tight and clear. Long reverbs and tap delays can very quickly get messy and get in the way of the vocal clarity.

Making the lead vocal stereo?

There is a proliferation of stereo imaging and doubler type plugins available today. These certainly add an extra excitement to the vocal, but remember also they communicate other things to the listener. They can ruin the sense of intimacy and naturalness. For certain genres, this fits with the aesthetic, but often this tool can be most impactful when used tastefully and intentionally to add difference and interest at different parts of the arrangement. Often this will be an intentional decision from the producer, so make sure that before changing things too drastically, you consult with them.

Arrangement

Even though this is not a choice for the mixer to make, it is important to realise the way in which stacking vocals is used to create different effects in contemporary music. Subtle doubles and vocal stacking can help to create a denser vocal sound. However, if this is not mixed effectively with sharp attention to detail it can end up sounding like mush and work against the desire to create an impressive vocal part. More detail on this is given in the section on tuning and timing.

Backing vocals

When we consider backing vocals, I refer to everything that isn't the lead. There might be different sub-categories of the backing vocals (BV) such as doubles, chorus stacked harmonies, ad libs, parallel lines, accents. All of these parts need to sit behind or wrap around the lead vocal with its job to enhance support or add interest to the lead vocal.

What are we hearing when we are listening?

- Are there any unnatural artefacts, clicks pops
- What shape is the sibilance in?
- Are all the transients/plosives aligned with the lead vocal?
- Does the BV compete with the lead, or swallow the lead?
- Does the BV take up a lot of sonic energy when it comes in in the mix
- Are the BV parts able to be differentiated, or are they blurred together?
- The space of the BV

Concepts we will cover in this section

- Cleaning it up
- Optimising individual layers
- Cohesion and blending
- Panning concepts
- Positioning relative to the lead
- Group vs individual

Clean it up and bounce it

We often start by also removing clicks and pops as we did with the lead vocal. Just remember that some of the popular tools for doing this involve AI process which means they are working very hard at analysing and separating these sounds in real time, so often they introduce noticeable latency. The best workflow to adopt is to bounce this processing once the pops are removed to your taste. This will be particularly crucial when working with BV's where there might be multiple individual layers that you want to apply this processing to which will increase the latency with each instance of the plugin.

Optimising the layers, clearing space

When we record layers of the same voice, it can have the result of building up of the tone of that person's sound as well as any room tone that is also captured. The layers can start to take up a lot of sonic energy where the buildup occurs that can compete with the lead and/or can have the effect that the individual layers (for example in three-part harmony) can lose their distinctness and start to sound like a blurry mud. There is no one way to do this, it is part of the fun and the creativity, the most important thing is to be aware and ensure you pay attention to it! If we have multiple part harmony, you might perform cuts to focus on a distinct range of the frequency. For example, the low part has strong low and low mid with high frequency shelf cut, middle removes the fundamental a bit and has some reduction to highs. High frequency clears away low frequency and allows the higher register to shine.

An important point to note: When we are working with stereo doubles, remember to balance the left and right by applying the same eq to both sides.

Panning approaches

There are a few different ways to approach panning. These are summarised below. The main thing to consider with panning is how you want the listener to experience the immersion. Do you want it to be realistic as if they were in a concert? Do you want it to cover and blanket the listener wrapped by the texture and harmonies, Do you want the parts to be blended or clearly separated in space?

LCR – this is as it sounds either hard left, hard right or in the centre.

Feather panning – with this technique you will position layers in a stereo pair in positions at increasing degree from the centre to hard panned. For example, three-part vocal harmony each doubled (so 6 stems), we can have part 1 at 90° hard panned, part 2 at 60° and part 3 at 30°).

Random – with this method you can create a more random separation of vocal parts in the stereo field. This might have the effect of a group of singers, rather than a clear uniform choir. With this method be careful to consider frequency balance across the stereo field.

Create cohesion

If the role of the backing vocals is to be heard as a cohesive group or layer, then compression on the subgroup of the vocals is often helpful to fuse the parts together and control the overall dynamic contour. A compressor like VCA is often a great choice on a vocal subgroup as it will encourage blending and cohesion through its 'glue' like characteristic.

Once we have individual spaces within the part we then need to start listening to it as a blended texture and position it relative to the other instruments in the arrangement. A 'character EQ' for tonal adjustments to the group can work well.

Keep the BV's out of the way of the lead!

A common issue with a highly developed vocal arrangement, is that the backing vocals start to dominate or cover the lead. It is important that they support the lead and don't make the lead seem small. There are a many ways that you can approach this.

Panning – leave the centre for the lead and position the backing vocals around the lead.

Imaging – you could use a stereo imaging tool that works with the mid and side.

Phase processes – you might chose to employ a phase or subtle time shift to delay the transient ever so slightly in order that the first sound to reach the ear is the lead and then followed by the backing vocals. One of my favourite tools for this is Soundtoys Microshift.

Fader – we often forget this, but you can turn down the group fader.

Reverb send – balance of wet and dry – more of the wet signal will push the backing vocal from the foreground and into the mid or background (depending on the style of reverb).

Equalisation – remembering that sound further away lose high frequency, we can employ that technique to our backing vocal subgroup to ensure that the lead vocal sits out in front.

It is also worth reviewing the section earlier in this chapter called 'fighting for attention'.

Group vs individual processing

You will see that in this section there were specific moments when I discussed processing the backing vocal layers individually and when I suggested that we would process the whole group. Whilst it is up to you to decide how you wish to organise your vocals, the principle applies to any individual stems that want to be heard as a group or as a layer. We would approach individual parts only when necessary and for the most part chose to approach these layers as a group.

For this reason, I often start with group processing for these types of multi-part textures and only go into the individual stems when I feel it necessary. For example, if I do not hear enough separation of the parts, or if I can hear artefacts that are particular to one track, or there is frequency build up.

Pitching and timing

Technically pitching and timing is not part of mixing, but sometimes you might decide it needs these adjustments. I chose to put this section at the end of the discussion on mixing vocals, because it is not technically part of mixing. I also have strong opinions and boundaries relating to mixing and vocal production. If I perform these functions, I become in part a vocal producer. This changes the nature of my role in the process, it may attract different fees and different royalty compensation.

Beyond considering the financial or professional issues around what constitutes mixing and production, there are also practical or creative reasons why I separate it from mixing. I feel that it takes me out of the mixing flow, I am focusing on something different than the mix. Once I realise it needs to happen I would do this outside or before the mix begins, I would then render out the changes and bring them into the mix file. Therefore, it is something that needs to happen before the mix begins. This workflow also has other benefits relating to the load on my mixing project and the general organisation and flow through it.

It is important to proceed very carefully when engaging with these types of adjustments to vocal parts.

If you mess with the timing, it can completely throw the whole performance and ruin the entire song. I would suggest that you would only make any alterations to a part if either you are hired to do so Remember less is more and only work on parts that need it (this is where your advanced critical listening skills come in handy!). If I felt that tuning was necessary, but I had not been booked for it, I might make some subtle adjustments and perhaps discuss this with the producer. If they don't notice it and like it, then that is great, but if they can detect that you have tuned something that they didn't ask you to it might create some friction and a lack of trust.

If I decide pitch adjustments are necessary, again unless it was explicitly said to me by the producer, I assume that any pitch adjustments need to be completely invisible, imperceptible, even the singer shouldn't notice it has happened. Ideally, they just say 'wow love what you did with my voice it sounds great!'. Singers know their voices very well, this means that you need highly advanced ears and a well-developed technique.

In my opinion, Melodyne is still the only tool that delivers seamless invisible adjustments for these types of situations. There are other faster methods, but these often a perceptible to those with keen ears.

Some tips on pitch correction

1. *Be very aware of the genre and the expectations of that genre*
   As I was talking about above, there will be a very different approach to this in a Justin Bieber style pop song than there would be for an acoustic folk singer. From that assessment, decide what approach to tuning is going to work best.
2. *Don't automatically tune everything perfectly in line with the grid*
   Unless it is a hard auto-tune song of course, which is likely part of the production and should already be in place. See below for more information on this point.
3. *Assess the character of the vocal*
   That might include the centre of the voice being slightly above or below 'perfect'. If you were to put the vocal automatically at 0 cents, this could really impact the quality of the vocal, and in some cases, the vocalist might say that it doesn't sound like them. Vocalists know their voice very well, better than you do (if you have just started working with it).

   Decide about whether the intention of the performance and the style of the singer is to be edgy or slightly ahead of the beat, or if they are loose and a bit late on the beat, does it fit the song? Was it performed this way on purpose.

   If you dive in and start adjusting without correctly assessing the above, the chances are high that you are making changes that are not in interest of the song.
4. *Develop a workflow that proceeds in short sections*
   The worst thing you can do is to have the vocal soloed and listen and perfect the vocal from start to finish. I understand the desire to do it, because it feels faster, but in the end it will not be faster because the likelihood that you are making bad decisions is high. We must always be constantly listening and assessing the effect.
5. *Listen listen listen*
   It should go without saying but it is so easy to forget. This can be managed by establishing a workflow that integrates working on small bits and then listening back.

6   *Broad strokes first and spiral into details only as needed*
    By approaching it first as broad and only adjusting the things that really stick out, you can be more confident that you are keeping the intention of the song and performance intact.

    When you work in sections or phrases, you want to take in the whole phrase. This allows you together determines what needs adjusting.

    As you get into more detailed tuning, ensure that you stop what you are doing and listen to these adjustments in context with pre-roll and the whole song playing to assess the improvement.

7   *Can you do more than pitch and timing?*
    If you work with an advanced tool like Melodyne, you might find that it starts to be a one stop shop for preparing your vocals for mixing. In my case, it ends up being faster and more efficient for me in addition to working on the pitch and the timing to also balance the loudness, as well as sibilance and get the envelopes for longs notes cleaned up as well.

Chapter 19

Take me on a journey!

The vocals might be perfect, the bass and kick are working the harmony is lush - but still there is something missing? If we don't create a sense of excitement or connection with the music, the most beautifully tuned vocals won't matter at all. The final finessing that we need to consider is paying attention to how the piece unfolds over time. Remember that we are involved a time based art and carefully pay attention to how this time is utilised.

What are we hearing when we are listening?
- The mix doesn't flow well from start to finish
- The mix doesn't build a sense of tension and release
- The transition from one section to the next lacks impact
- The mix doesn't hit its peak at the right time, or at all
- After the song is finished I don't feel compelled to play it again

Concepts discussed in this section
- Arrangement
- Loudness map over time
- Frequency
- Automation
- Performance and imperfection

Don't start too high

When we start to consider the piece from start to finish, we need to remember to build the momentum. Paying attention to the opening bars is vital. It needs to have enough energy to captivate the listener quickly, but it also needs to keep some in reserve so that the track is able to maintain that interest.

You might be considering that this should fall under the discussion of production and it is not a mixing decision. This is true of course, but when we get to the mix we have the final opportunity to check if and how this is happening. We can accentuate the differences through clever use of filtering as well as fader automations to create movement and difference between the sections.

Composing the joins

We hope that the producer has been paying close attention to this before it gets to the mix stage, but it is often something that we might need to also pay attention to when mixing. It is our job to ensure that the transitions are hitting with the right energy creating a sense of flow and energy in the right places.

When considering the joins, the transition points, we consider the momentum of the song and how these transitional points contribute to the momentum. We might consider the levels, but we also might work with filtering, subtle processing and automations to unnoticeably remove some of the audio spectra before an important impact to help that moment reach its potential.

Golden mean

The golden mean is a ratio that abounds in the natural world. It is sometimes called the divine mean. It relates to the Fibonacci series. In philosophy, it is often about finding a medium point between two extremes. So why am I talking about it here? Well we often consider the golden mean in relation to the balancing of the piece. When considering the golden mean in music, we might want that the highest point of the piece is at or near the golden mean point. What is the golden mean? It is at two-thirds, or .666. So by this concept, in a pop song the highest point is there – which often means the final chorus. If we have the highest point too early, the end of the song is boring; if we have it too late, we might lose our audience because they lose interest. Whilst we are not producing to the golden mean, we can ensure that we create the highlight through the mix, by adjusting the elements so that we create the most energy at or around the golden mean. Try it out if you haven't before!

Subtle changes and automations

Often in a production there are great little details within the arrangement. At this point, we can employ simple fader automations, to highlight different elements at different moments to direct the listeners attention through the arrangement.

Don't be too perfect!

We are often taking so much care to perfect everything and fix all the mistakes that we forget that we sometimes need sounds to stick out, or to interrupt us. If everything is perfectly balanced, then we run this risk of making everything blend. Remember too much of anything is not good, our job is to balance the elements and this also means blending elements with allowing individual elements to stick out and give shape and difference.

Part Four

Important next considerations

Summary

If you have made it to this part of the book, you should be feeling confident and clear about your path to continued development as a mix engineer, and you should have some quality mixes under your belt as well.

In the next section, we talk about other details that are important to consider in relation to mixing. Often these are questions that students ask me as they develop their practice. I have put this discussion in a separate section, because I believe they are concepts that you can deal with as a second step after first focusing purely on mixing. These aspects are not things that you should really consider too much when you first start mixing – you have a lot of other things to understand first. Once you have digested the first three parts of the book and assimilated the various concepts into your mixing, you can start to consider the details discussed in the following pages.

Another important note: it is not wise to go through this section if the earlier parts of the book are still unclear to you.

Chapter 20

Preparing for a mix – moving from production to mixing

Preparing to mix, moving from production to mixing is something that causes many producers confusion. The difficulty with knowing what and how to bounce out their project can also lead to the decision to mix within the production. This is fine of course until as long as your cpu can handle it, however if you outsource your mixing one day then, the difficulty arises. This section considers if you are mixing your own productions, if you are sending your mixes to an external mixer, or if you are the external mixer what you can/should ask for.

Mixing your own work

Should I mix in my production or not?

My perspective traditionally was that it is better to mix with a clear mindset, in a fresh project. As music evolves and the style of different genres blurs the lines between mixing and production I don't think that it is always as cut and dry as that anymore. When possible, I still think it is beneficial to start a mix clear of the production.

How do I prepare my production for mixing?

If you want to transition out of your production and into a mix, there are two general approaches:

1. You bounce out stems of your tracks and import them into a new project
2. Rename your production file, flatten processing in your production and tidy it up

Either way the key realisation here is that you are making some decisions. This is why I feel that this moment is very important. Decisions are everything, this is how we build our aesthetic. One of the big issues I observe with people that are 'mixing while producing' is that they struggle to make decisions, and end up second guessing both their production decisions and their mixing decisions and the final result suffers.

Tips for bouncing out your stems/flattening tracks

- **Make a plan** before you bounce: By this, I mean consider the sounds. Are their sounds that want to be heard as a group? Have you spent time blending and creating

a texture/layer that combines multiple sounds in a specific way? If so – these should be bounced as a group, i.e. one stereo file of the group
- **Separate your low-frequency sounds.** These are the ones that often have issues, and if they are bounced in a group with other sounds, it can make it more difficult when trying to fix issues in the mix. So this means, a kick track separate from the other drums. Kick and bass bounce separately also. It doesn't necessarily mean that you need to bounce all four of your kicks and three bass sounds as individual tracks (remember the above point)
- Sounds that perform the same task (i.e. are fx and transition sounds) not overlapping bounce together, they can be easily separated later if necessary
- Bounce tracks **without reverb and delay**. If in doubt, bounce one dry, one wet of the same track so both options are there
- **Bounce parallel channels** with clear name, e.g., what fx – if you want the mixer to honour these choices in the mix
- **Consider routing and side-chain** signals – check that your side-chain processing has exported the way you intended
- **Bounce lead vocal separately**
- **Leave on processing** (apart from reverb/delay)
- To turn off or leave on? If you turn off the processing the sound changes dramatically, then it means that your processing has built a specific sonic aesthetic and should be left on
- **Turn off mix bus processing/limiting on the master** – if your master buss is doing a lot of work, then this point might need to be ignored – actually what you would likely do in this case is the 2nd option of converting your project into a mix project and flatten the individual tracks and keep the processing on the master bus as you go into your mix, where you can tweak and adjust what you have already done on your mix

You can always go back and bounce another stem! If you are mixing your own work, you can easily open up the production file. The point of this process is to make some decisions, which helps you to move forward in the process of creation. These decisions lead to a fresh perspective and some more mental space to focus on the mix. It cleans up the visual workspace. Countless times when I mentor mixers we get lost looking through their processing which includes endless plugins. It also has the benefit of allowing you to gain stage with ease. Remember go with your gut, be bold and commit those crazy ideas, remembering that you can always go back into the production file and turn it off and bounce it out if later you realise it is not working.

Chapter 21

The role of a mix engineer in the collaborative process

This book opened with a discussion clarifying what mixing is. It was important because it is something that can be unclear for many people starting out. Even beginner mix engineers are often not fully aware of the role of a mix engineer, and how they fit into the overall process of building a commercial production. This lack of understanding is often more confused when a producer mixes their own work. Clarity about steps in song production will optimise creativity at every stage and will ensure that best focus and efforts are applied leading to a better result. Each step of creation has a particular energy and focus. If not attended to, the focus is not directed in the correct way and the ultimate success of the piece suffers.

Different stages different focus

1. **Ideas** – concept, lyrics, chord progression, melody, groove, specific sound
2. **Songwriting/arrangement** – create a linear progression from beginning to end. Build out the ideas, create a cohesive structure and form
3. **Production** – expand and/or refine the arrangement, develop the sonic identity through sound design

    - - - Here we cross over from the creative and into the technical - - -

4. **Mixing** – optimise the sounds individually and collectively, solidify the aesthetic, ensure the momentum of the piece is as it should be, that it fits with the desired genres, etc.
5. **Mastering** – Final adjustments of the overall sonic quality of the track, loudness, tone – prepare for distribution

Being mindful of these distinctions is important when working on a production from start to finish. We don't want to be pre-occupied thinking about the details of mixing whilst we are producing, it will take us out of the flow by getting lost and not focussing on what we should be. For example, if we spend one hour refining the relationship between the kick and the bass as a mix engineer would, we have taken ourselves out of the flow for designing the arrangement and the overall aesthetic whilst producing. It doesn't mean that when producing you can't or shouldn't spend any time refining sounds but just that we keep in mind that the time to perfect this is coming later in the next step and to stay on course working on the production. Once you have found the correct bass sound and a kick tone that works with the bass and arrangement, the rest can be left to the mixing stage. Leave the deep analytical attention for a later point, and staying in the flow of sonic aesthetic and shaping the overall track as much as possible. By knowing the key decisions that need to be made in each step in the overall process, it allows you focus your attention on elements that needs attention from the right perspective at the right time where to spend your creative energy. You can't mix the track before the arrangement is finished, just as you can't record the vocals until the lyrics are done!

Not changing, but optimising

Above I separated the processes of songwriting and production from mixing and mastering. The first three I classify as creative and the final two are more technical processes. This is not to suggest that mixing and mastering do not involve creativity, but in the sense of 'creating the song' – it should be fully formed before it gets to the mixing stage. Our role as mix engineers, is not to change anything essential, but to adjust, optimise. We work always in service of the song first, and we discuss any changes that will impact the aesthetic and arrangement with the producers/songwriters. Being aware of our role in the collaboration will assure that we do good mixes and we build good trusting relationships with our clients.

Chapter 22

Psychoacoustics

Most people have heard of the term psychoacoustics, but many people don't really know what it means. There is a general lack of understanding about the impact that various psychoacoustic principles can have on the mixing process, as well as how psychoacoustics are at play when listening. Whether you know it or not as a mixer you are highly engaged with psychoacoustic phenomena!

What is psychoacoustics?

It is the study of how humans *perceive* sound but it also includes physical properties of sound and the impact of these sounds on human's hearing apparatus.

How do psychoacoustic concepts relate to mixing?

Psychoacoustics are important to remember on two levels for mix engineers. Firstly, we need to be aware of certain concepts to understand the limits and the influences that our brains are under when listening. Secondly, with knowledge of psychoacoustics, we can utilise these phenomena to manipulate and craft specific sonic experiences for our listeners.

Noise exposure and hearing sensitivity

The recommendation from the "Standards for 'Safe Listening': Past, Present, and Future" (Kawamori, Best, Laureyns (2020) is that we can listen at a level of 85 dB for 8 hours without any damage. So this should be set as your standard mix maximum level. If you want to have some louder moments, 90 dB is the maximum you should go for longer session, 2 hours maximum. We can be exposed to 100 dB for 15 minutes without any damage. Now I'm not suggesting that we mix this loud, but present this now to consider whether you use hearing protection when you are out at a gig or at a club? These levels often exceed 100 dB and if you are not wearing hearing protection as someone with a career in sound, you are putting yourself at a disadvantage.

The acoustic reflex – dulling effect of sounds after exposure

> This effect, combined with the effects of temporary threshold shifts, can result in a sound level increase spiral in which there is a tendency to increase the sound level "to

hear it better," which results in further dulling and so on. The only real solution is to avoid the loud sounds in the first place. However, if this situation does occur, a rest away from the excessive noise will allow some sensitivity to return.

Loudness perception

Below is a picture depicting the Fletcher Munson's curve (Figure 22.1). This graph shows sensitivity to loudness at different frequency bands. This tells us something important about how tuned in we are to different frequencies. This curve is particular to humans. If we were analysing the same thing for a whale or a dog, it would be different. Unsurprisingly we have a great sensitivity around the 1–4 kHz range which is where we make sense of speech. It is the home of all the transient and consonant sounds of language that gives the vowels meaning forming words.

If we look closer, we can see that the graph also tells us that as the volume gets louder the curve flattens out. So that tells us that at loud volumes we have the most flat and 'true' perception across the frequency range. Does that then mean that we should mix very loud? In theory yes, but in practice there are other aspects at play which mean this concept doesn't work. Above we talked about the acoustic reflex and the way that our sensitivity to sounds dulls after exposure. So how can we build a listening habit that is healthy, will avoid the acoustic reflex but will also consider the Fletcher Munson's curve? The answer is in maintaining a good level and having breaks. My approach is to work at around 85 dB maximum of two hours. Incorporate times when I reduce the level for quiet attention and only pump the volume to get a vibe in very short bursts. The key thing for me is to take a break every two hours. This naturally resets my ears and they have grown accustomed to wanting the level 85 dB.

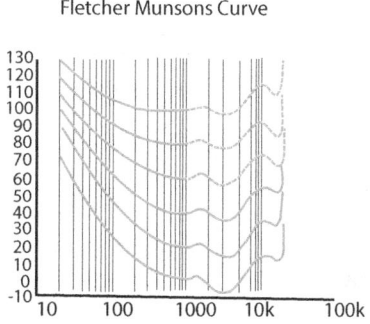

Figure 22.1 Fletcher Munson's curve

Just noticeable difference

This concept discusses the smallest changes that listeners can notice. This impact us at many stages of the mixing process.

Volume – the JND for loudness is typically **1–2 dB**. So if you change something less than this it will likely not be perceived.

EQ – this is an extension of the above point, but also now keeping in mind the Fletcher Munson's curve and the fact that our sensitivity changes depending on the frequency. To in regions where we are sensitive, i.e. in the mid-range we will be able to perceive more subtle changes, but in the extreme low and high we will not be as sensitive.

Compression – works with amplitude so of course the JND comes into play here. Once we break that 1–2 dB threshold we become much more sensitive to adjustments in amplitude. This is even more noticeable when a compressor is placed on a subgroup/bus.

Critical bands

This concept considers the ways that humans process a single auditory event and the range of frequencies that we hear as a unit. Within each critical band sounds overlap. Build-up of frequencies can cause masking and interference. The bands change based on the central frequency. Lower bands are wider and upper bands are narrower. This concept can be seen at play in the design of many equalisers. I often use an exercise with mixing students to get them to categorise the frequency range into these broad critical bands including sub, low, low mid, high mid and then presence and brilliance for high bands. (See equalisation in Part Two.) In practice in the mix, having an awareness of this can help diagnose issues with masking and consider either arrangement changes when two sounds occupy the same place, or in a mix by utilising equalisation to shift our focus to different area, or to reduce the presence of one sound in an overlapping band.

Precedence effect and Haas effect

The precedence effect in psychoacoustics relates to the way that the brain determines the direction of a sound source. According to this concept the brain determines sound localisation based on the first sound to hit the ears. Later sounds (called the first reflections) are used to create a sense of the space that we are in (this is how virtual space algorithms are built). However, there is a timeframe whereby the human brain cannot detect differences, this is between 1 ms and 30 ms. This is known as the law of the first wavefront or commonly as the Hass effect. The brain fuses together multiple sounds that occur within the window of the first 30 ms.

Directional illusion

The whole concept of stereo mixing was created by manipulating psychoacoustic effects to fool the ear into perceiving sounds positioned in space. This concept covers so many aspects of mixing in stereo. It relates to the concept of panning and stereo imaging, the use of phase processes, working with spatial processing. This gets even more complex when we move into the world of 3D and immersive mixing where we are working with HRTF.

Inter-aural time and intensity difference

The average distance between a person's ears is 18 cm; thus, the sound will take a different amount of time to travel to each ear. But it doesn't just take longer in a straight line, it has to travel around the head, as well. Then there is also the shading effect of the head's mass and constitution. This has an impact on the frequency and the amplitude of the sound when it reaches the ear furthest from the sound source. These ITD and IID differences also relate to mixing techniques and in fact the entire concept of stereophonic listening that we all take for granted. 650 Hz approximately the threshold for the heads ability to scatter sound waves. This tells us important information about the part of a sound that helps us localise sound in space. Often the front part of the sound, or the transient has information above this region.

Mixing is psychoacoustics!

The art of mixing is all about working with the phenomena of psychoacoustic to create a perception of an experience – the perception of loudness, the perception of space and depth, the perception of movement, etc. All of these things are somewhat an illusion, and part of the magic of the sonic arts. Remembering that mixing is creating virtual worlds and virtual experiences always maintains my wonder and reverence for the artform!

Chapter 23

Aesthetics and individual mixing style

What are mix aesthetics? How are they different from the overall sonic aesthetic? Is a mix aesthetic different from production aesthetic?

This topic might feel a bit strange, particularly following the discussion on collaboration and identifying the role of the mix engineer in the overall process. I pointed out that the role of a mix engineer should not be building the sonic aesthetic rather focussed on enhancing and optimising. This might make us consider then that mixing is entirely technical and the changes should be as invisible as possible. Well yes, and no. This is where it can get tricky, but stay with me for a minute. All of the tasks that we undertake are creative, depending on where we are in the process these creative choices, will have more or less obvious impact on the overall song. To give a simple example, the choice of chord progression and melody are really obvious whereas the choice to use tape saturation on the mix bus is less noticeable. Both will impact the overall aesthetic of the song to greater or lesser extent. Even though mix engineers are not creating the sonic aesthetic from the ground up, they contribute to it through their unique style of mixing. Every individual will have their own style which is developed by the things they like and don't like. Some mixers love analog sound and tape saturation, whilst others prefer clean digital processing. Others like it loud and boisterous whilst others like clean and refined. This is where knowing your own mixing aesthetic is important. This becomes your own style that helps you stand apart from other mixers.

As a mix engineer we acknowledge that the approach that we take in enhancing and optimising the production will also leave a footprint. Therefore, we focus on ensuring that our approach is aligned with the intentions of the song and the production. For this reason, you might find that not every type of music production fits your mixing style. That is ok. You can try to mix everything, and perhaps in the beginning you will take on every job, but after a while when your personal mix aesthetic starts to emerge you will realise which productions your mixing thrives with. Once you know your own mix aesthetic it will help you make smart choices when you are offered work, aligning with the right productions will help the music flourish and demonstrate your skills in the best light.

When we consider big name celebrity mixers, like Scheps, Lorde Alge, Braithwaite etc. Each of them has a signature style, with a very specific approach, that they have managed to package and brand. This has lead to other mix engineers emulating their style and is now to the point where people actually want the mix aesthetic to colour the production and overall sonic aesthetic.

Ultimately it is important to work the way that feels interesting and natural to you. If you try to please everyone, you will likely end up pleasing no-one. Get excited about your mixing and develop your own style, it will help you carve out a path in your career. But remember to pay attention that the mix aesthetic does not change or detract from the sonic work's own character. If your style doesn't fit with the song, better not to force it – for everyone's sake! This can be one of the hardest lessons to learn when you are trying to build your career with the sense that you need to take on every project that comes your way. As creative people we should only work on projects that we are excited about. This will ensure that great music gets made.

Saying goodbye to references?

When you are learning and developing your skills, your ears, your knowledge and your confidence as well as when you are working in less-than-ideal conditions for mixing, references are a real help for a multitude of reasons that I have already detailed in Part Two of the book. Parallel to this point, following on from the conversation about developing your own style, is there comes a time when the apprentice must take flight and forge their own path. So the question for this section is when can I say goodbye to references? When are their times that I will still want to use them?

There are no absolutes. As always everyone approaches this in their own way, it is a creative pursuit after all. The decision to work less with references comes when you can identify your style emerging in your mixing work. It comes when you are confident that you can get a quality mixing result, you have a keen awareness of the low-end balance that you are after, you are sure that you can get the vocals to sit in the right place. Knowing when to leave references aside can be difficult, but when you have people calling you to mix after they have heard your previous mixes and they like how your mixes sound, you might say ok, yes I have a sound that people want. Perhaps you mix your own productions and have developed a style that is being well received, or that you feel really satisfied with – in this case it is also probably time to stand on your own feet. All of these moments tell you that you've got this! And to move forward confidently.

It is still always a good idea to keep in mind that you might dip in and out of using references in your work. Maybe there is a new style that emerges that treats audio in a new way and you want to get a handle on it – by all means get your analytical hat on and deep dive into it. You might also find yourself working with clients that have strong ideas what they want, but they lack the ability to describe it in words – maybe they can pick out songs that they love and point to the way the vocals sound or the treatment of the low end for example. In these cases, it is really important that you as a mix engineer have the skills to extract the information that they are pointing to in reference tracks to give them the result they are after.

Chapter 24

Mixing whilst producing

Most of this book approaches mixing as a separate process from production. It is presented as a process that happens independently after the production is finished. This is the traditional approach for commercial work since 70s and dominates to this day in commercial settings. Mixing was not always such a separate part. When we consider the beginnings of recorded music in the 1950s and 1960s the limitations of recording media meant that all the 'mixing' had to be done before or during the recording process. In the recording studio, there were innovative approaches to positioning within a shared space relative to the microphone that was capturing the sound made in the room. There was also a great deal of attention given to the art of orchestration and arrangement. I say all of this to remind us all that nothing is set in stone. As a creative process we are constantly negotiating the tools that we use, styles and trends, experimenting with approaches and so on. So whilst mixing has emerged as a separate from production it is not always the case. In this section, we consider why it is helpful to separate mixing and production, before looking at the types of music that can benefit from mixing whilst producing. We finally look at best practice or some considerations that are helpful to ensure that nothing is being missed when combining multiple components together.

The argument for separating mixing and producing

The first point to raise here is that mixing and producing employ a different type of mental engagement and attention. I always use the example of two approaches to working with a common effect process. Let's take chorus. Using this with a production, sound design aesthetic we might work with chorus to change the sound of the guitar to give it a particular quality. When using chorus in mixing, we don't want it to be perceived as changing the quality of the sound, rather we might use it to subtly introduce time/phase differences to help us position the sounds for the listener. Same tool very different goal. One of the issues that I see with those who mix whilst they produce is that they often remain in the production mindset and think that they are mixing when in actual fact they have not really started it yet. If you are the kind of person that wants to mix whilst you produce, I would highly recommend at the least practicing mixing and critical listening as a separate process from production as you are developing your skill.

Another argument for separating the process is about mental space and focus. By bouncing out stems and starting with a clean slate, you can focus on what is working in the mix and where you might need to make adjustments. Working on mixing within a heavily stacked production project can not only tax the computer power but also can be taxing for your mental focus.

The argument for combing mixing and producing

In certain genres, the line between mixing and production starts to blur. It can be difficult to know where one ends and the other begins. A genre that fits into this is techno. Often in techno the approach to processing (for example dynamics processing) is part of the mixing but this also shapes the production. In genres like techno, there is also often complex routing and side chaining where elements impact and control facets of the processing of other elements. This works for the style but often also serves important function from a mixing perspective. When we consider more experimental genres such as electroacoustic music, the line between composition, sound design and mixing is even more blurry. The more experimental the music the likelihood that the clear boundaries and borders are harder to define. If you feel that your genres or just your personal style is to approach mixing and production together, then go for it. However, this doesn't mean that you should not still practice mixing as a separate skill that you need to learn. By all means, combine the skills, but remember that first you *need to develop the skills.*

Some considerations when mixing whilst producing

Mixing still has to happen!

Mixing and production involve different mindsets whilst working with the same processes. It is no wonder it can be a challenge to keep it clear in our heads and to ensure that we are paying the right amount of attention to each process. Production is about defining a sound, creating a sonic identity and locking in the arrangement and the core elements of the music. When we get to the mixing stage, it is about tightening the elements and ensuring that it is all working as it should as we have gone to pains to present in this book. When you mix whilst producing there will be a need to juggle the concerns of both elements together. Often beginners who mix whilst they produce end up focusing on the production and the detailed attention to the mix gets forgotten. Don't let that be you! Even if you would like to mix and produce, when you are learning it is a great idea to practice mixing as a separate process from producing. This way when it comes time to combine the tasks in your own productions you have developed the skills.

Mixing, orchestration and arrangement different sides of the same coin

If you are blending mixing and production, it is a great opportunity to take advantage of the ability to make adjustments to the arrangement when you come across issues in the mix. When there is frequency masking – you might ask yourself, could I re-voice this part? Could I change the sample so that the quality is not so close to another sound that it is clashing with? You also have the ability to turn things off that are not working. Rather than slamming everything with compression and side chaining to try and force it all to work together it you might realise that you should first tighten up rhythms, registers and consider the relationship between the parts leaving space and enhancing interplay between them.

Be aware of adding processing on top of processing

The likelihood of losing your focus and just adding more on top of what has been done is high. As you progress through the production and integrate mixing, it is important that you make decisions. The benefit of mixing whilst you are producing is that you can adjust the processing rather than stacking it up endlessly. It can help to re-centre the central role of mixing as critical listening so that it is not just about processing but evaluating what is there and making decisions about what is working best. Be mindful that you are not just trying new things out but making decisions.

Flatten and commit

Decisions are key. One way to keep a sense of progression from a decision making perspective is to flatten stems that are finalised. Even though you are working within a production don't be scared to flatten elements that you have finalised processing. This helps you to create mental space as well as clear your DAW and relieve your computer DSP. Always save the adjustments under a new project name so that if there are any issues, it is easy to come back to the production pre flattening.

Putting off decisions?

In many of the points made here, the word decision comes up. Making decisions will always be fundamental, this is how we create, every small decision adds up and that combination creates our aesthetic. Putting off decisions means that you don't know what you like or you don't have the skill to create what you want. Both issues can be approached once diagnosed. Be aware when mixing whilst producing if you are falling into the trap of putting off decisions.

Develop a workflow

We talk about workflow in this book relating to the mixing process, but if you are combining production and mixing, create your own bespoke combined workflow. This will help you to be sure that you are not missing out on anything. Keeping your own checks and balances is key to confidence in your own process, whatever that process is.

Mixing in a genre bending landscape: towards experimentalism

This book is about mixing sound. There is at times a leaning on the popular song format, as much of the commercial work for mix engineers deals with this format. But it is by no means limited to working with this format. Different genres of music move away from the song format, and whilst they might have some specific features that need special attention, they are still sound and considering the way that sounds combine together to create a wholistic sonic experience. Dance music genres, are not songs, they are often repetitive, and progressive in form, they often don't have lead vocal lines, in these formats, movement are fundamental and dynamic motion and processing are a great focus. Conversely perhaps you are working with electroacoustic music, or ambient sounds, these types of sound productions will lean more on concepts around space and textural blending and transformations.

The approach that has been employed in this book is to develop an awareness of what you are working with as a mixer, to analyse the project and to be able to select the right tools and approach that will yield the results you are after. As new music is created, the more experimental styles find their way into the mainstream and we introduce greater complexity into our sonic creations. The boundaries continue to blur. That fine, with a good grounding in all the skills you will be ready to metamorphose your approach with the sounds that you create.

Chapter 25

Mix bus processing

Mix bus processing has become so common that it almost seems like the standard way to mix these days. Is it inherent in the process? Or some new trend? This section will discuss the evolution of mix bus processing. Discuss the concept of top-down mixing and of course also show how to do it and offer some tips for best practice.

Evolution of mix bus processing

In the days of the first mixes, it was all analog. So we can trace mix bus processing to the evolution of mixing as an analog format. Originally mixing was minimal, with the first 'mixed' records having only a four track console to work with there was a great deal of mixing occurring throughout the recording phase. One famous early example of this mixing/consolidating parts as you go is the 'wall of sound' created by Phil Spector. These records were using four track technology and the recording process involved bouncing down recorded layers onto one track and layering. The very act of processing to four track tape and then mixing down to one assumes a level of associated mix bus type processing, through the saturation qualities inherent in tape (see Part Two on saturation for more clarification).

As we move through time and audio technological development many innovations were made. One of course was the development of large format consoles as well as external compressors and limiters. We moved from four to eight track mixing in the 60s to large formats in the 70s. Neve's 8028 was the first large format console that included the ability to process on the master bus. The console was equipped with advanced routing and ability to patch in external processing units. It was common to utilise a neve 33609 or a Fairchild 670 tube compressor on the mix bus. The processing style during this period was warm, tubes, gentle approach to processing.

It was not until the 1970s with the introduction of the SSL 4000 series consoles that we first saw a VCA compressor built into the mixing console. This compressor changed the game and became a world standard. If you have ever used a 'glue' compressor emulation of any type, it is inspired by the SSL mix bus compressor. The SSL VCA compressor sound creates a 'glued' cohesiveness to the mix, it is quite responsive and it can create a famous 'punch' and energy to the mix as well. The characteristic of this sound is more up front and powerful.

Once we get into the 90s and move into the DAW environment we can see the exponential growth traced to today of plugins. Now we have every option at our finger tips, so it can be overwhelming to know what to choose.

Top-down mixing

This approach favours considering the mix as a whole. So rather than starting the mix building it from the ground and adding and expanding the mix, the top-down approach considers the overall balance of the mix and starts first with processing the master. It only approaches individual tracks if needed after first taking time to really tune in the sound and process on the master bus. Common processing will include compression, equalisation and use of saturation on the master bus.

How to set up your mix bus

The first step with mix bus processing is selecting your tools. With the immensity of tools available in the contemporary digital environment it can certainly be overwhelming. When getting started with any new approach in mixing, my suggestion is always to do your research and learn from the productions that you love the most. Those sounds that you are inspired from can be a great source of inspiration at any level. Find out who mixed the track and investigate their approach to mixing, which hopefully will include information about how they process their mix bus.

Another way that you might find your way through the maze of options is to consider how your sound relates to the eras of music. Does it have a vintage feel like the 60s or 70s does it have a more in your face energy of the 80s and 90s is it very digital and clean and transparent. Each of these approaches can be paired with processing technology.

My standard bus chain is as follows:
VCA Bus compressor → Passive EQ → Master Tape → Meters/Limiter for mix checks
SSL G Bus → Manley Massive Passive → Ampex 2 track → VU, Correlation, Peak

Note: whilst I always use the SSL Bus compressor as standard. I might not always use the EQ and the tape I might also turn off if I'm working on music that needs a real sharpness to the transient material

Good practice tips

Always MIX INTO the processing. That means set it up first. You will notice in the 'set-up' section of the book between Parts Two and Three the order of set-up includes when to set up the mix bus. This is important! If you put the processing on the end, this is not mixing into a mix bus, but more like mastering.

Gainstage the mix BEFORE you put your mix bus on. This is particularly important for mixing into processing like compression.

Don't mix into a Limiter. It is ok to set up limiter on your mix bus, but have it turned off and only use it to check once the mix is done. This is helpful for checking what the mix will sound like after mastering and for bouncing for clients, but mixing into a limiter is not generally recommended.

Be aware of the way that the mix bus processing impacts the Low end. You might like to use a compressor with a side chain filter on the lo end.

Remember to be gentle. Particularly for EQ and select one that favours smooth tonal adjustments.

Consider your phase and power: Is the mix bus adding a little bit extra? Or is it squashing or diffusing the mix in some way? This means selecting the right tools. So be sure to use an EQ that does not create unwanted diffusion through phase distortions.

Chapter 26

Loudness

As the ambient noise level that most of us experience in our day-to-day lives gets increasingly louder, the music that we make has tried to keep up against the racket. When we listen to music made 60 years ago (if that haven't already remastered it to make it louder) is much quieter than today. The quest for loudness is one that many producers, mixers artists and labels are concerned with. In this section, we open up with a consideration of the loudness wars before going on to discuss aspects like genre that might determine where you should stand on this topic before going into a practical discussion about some consideration to help optimise the loudness in your music should you wish to do so.

The loudness wars

Have you heard of the loudness wars? Humans generally perceive something louder as better, or more impressive and when we hear two songs side by side, the one that is louder will generally be considered 'better' (I don't necessarily agree with this, I'm just explaining the loudness wars concept). In the competitive world of commercial music, this was exploited by labels and artists alike in the bid to make themselves and their music stand out from the pack. It hit a peak in the early 2000s where the labels were competing for radio time and pushing the mastering engineer to get the song louder and louder. But what happens when we have reached the maximum? Where do we go from there? Well in a sense… down… we start to clip the loud parts of the signal creating harsh distortion and continue to bring up the lower dynamics within the track – the result is a perceptively louder song. But at what cost? Some of the tracks in this period seem to have introduced so much distortion that they are very hard to listen to. Things were getting out of hand.

Fast forward to the development of streaming and someone had a great idea *what if we introduce a standard that songs streamed on these platforms will be normalised against*. This lead to the introduction of LUFS and supposedly when LUFS were introduced we could finally lay down our clippers and our limiters, breathe a sigh of relief that the battle was over, we could go back to dynamic listening and save our ears. The nature of humans is that we find ways to tweak and to compete nonetheless, so the loudness wars in some senses wages on.

Genre and loudness

Not all music benefits from being made to sound loud(er). Certain music needs movement and dynamics. We must remember that we don't get something for nothing. In order to increase the perceived loudness, it means a reduction in dynamic range.

For certain genres like classical music and jazz, it doesn't make sense to make it too loud as it goes against the style. Most people are aware of this, but don't often think about it in other contexts. When we move into the spatial environment, mixing Dolby Atmos, etc., with the same mindset for approaching a loud stereo mix will produce terrible results, that will ultimately cancel out the spatial aspects. Conversely, there are genres where maximum loudness/lower dynamic range is part of the style – one example is techno, and I'm sure those of you reading this have your own additions to this category.

Remembering our role as a mix engineer, we must consider the song and work to the intentions that the song reveals and discuss these important elements with the clients that you are working with.

Always remember that you don't get something for nothing – and be mindful of the relationship between loudness and dynamic range and movement.

Loudness starts before the mix

Many times I have received productions to mix with clients asking me to make it loud and powerful and I have to manage the expectations. It starts with the sounds themselves – how they have been designed their qualities.

Then, it relates largely to the arrangement – if you have lots of sounds trying to fight through in the same frequency range it will likely have a negative impact on overall loudness. The best arrangements that are well orchestrated lead towards loudness that maintains depth.

Loudness optimisation during mixing

Of course, there are things we can do during the mix – these have been talked about in detail throughout the book.

If you need a refresher, you might go through the book paying attention to key words dynamics, envelope, transient, resonance.

Remember that equalisation can create unwanted phase and can weaken your signal, so when you are working with EQ be mindful of that and consider the overall impact. Also remember that you don't get something for nothing and as you approach dynamics pay attention to how the dynamic reduction impacts the sound in question and in relation to other sounds.

Loudness and the mix bus/master

When working with young producers who are getting into mixing often one of their main concerns is that their music doesn't sound loud enough when compared to other commercial releases. Whilst there is certainly much that we do in the mix to improve the perceived loudness by working with EQ to clear space when instruments are masking other sounds, as well as working with compression to control the sound and things like saturation to introduce harmonics to support perception of tones.

When I am mixing to optimise loudness I always want to have a good sense of how loud my mix is, so monitoring is important. How are you monitoring? are you aware of the average loudness? Do you have goals for this? Do you have a specific average loudness in mind that you want to achieve during the mix? A more dynamic mix will have an average loudness of something around the −18 dB RMS, and a techno mix might already have a loudness level of −10 dB RMS. There are no hard and fast rules, what matters is that you have your own approach and you know what you are doing. Mixing with monitoring will also let you know when there are sounds occurring that are taking up a lot of energy in the mix, these sounds will blow out the master level, and is a good sign that these sounds are either being masked or masking other sounds so you can work your eq and compression here to control these and get them to fit in the mix without taking up all the space. This will give you more headroom which in turn means you master will be able to be louder.

When I'm working on louder styles, I often find that my mix bus is doing more work. I often have in addition to my standard bus compression, saturation, maybe some multi band dynamics. You might like to review the mix bus section.

Mixing into a limiter is generally not recommended, but checking a mix with a limiter is common practice.

Loudness and mastering for a mixer

This book is not a mastering book, so I don't intend to go into detail about mastering. The reality is that some of us will do masters as mix engineers, and so it is helpful to at least talk about some basics. I preface this section by saying that any commercial release I do I prefer to send it to a mastering engineer, for my own more experimental releases often I am happy to manage the mastering myself. Below I will talk about the types of things I think about when I'm doing my own mastering. For full training on mastering, dig deeper into specialist mastering resources.

The first rule to remember is that mastering is not a magic pill. There are some amazing things a good master can do, but if it isn't right in the mix it won't magically be fixed in a master. So in the context of loudness if you want a loud track, you need to consider all the previous points, rather than thinking that slapping a limiter, saturation and clipping during mastering will work. It will make it louder, but it will most likely sacrifice the fidelity of the music that you crafted.

The second rule is be aware of resonances that might be taking up energy that don't need to be there. Now when you are going to master be careful what EQ you go for, consider the relationship that phase in EQ might have on the mix – the wrong approach to EQ can actually weaken your mix rather than make it more powerful, so ensure that you understand this concept fully and consider linear phase for certain EQ tasks on the master that have surgical type shapes. I also tend to use passive EQ's like tube style, because of their minimal phase disturbances.

The third tip is to consider how you gain up your track – often people approach this by using multiple processors that turn up the level in different ways, so in other words utilise multiple stages of gain, rather than one dramatic limiter. It is helpful to explore the differences between different limiters and understand how they work and what they do to the sound.

The fourth tip is remember our good friend saturation. Again it is important to be very careful here and select the appropriate type of saturation and always be attentive to hearing what happens. Some unwanted artefacts of saturation can be low mid boost where bloating and muddiness can occur. They also tend to smear the transients, if we are doing an aggressive trap or techno mix we don't want this to happen.

The fifth tip for those of you wanting to have a go a optimising the loudness is to explore clipping. Clipping is the process of reducing the dynamic range abruptly, by capping or cutting off the loudest part. We should all know by now that the transient information is most often the loudest parts, and in certain genres like say techno or trap, we often have very loud kick, snare and sometimes hi hats. It can sometimes be helpful to bounce out a section of your mix and analyse it visually. If you see that the transient information is much louder than the average loudness, then you might be able to employ clipping technique to bring down the dynamic range of the transient information, thereby giving more headroom to turn up the mix overall, thereby increasing the average or perceived loudness. Different types of processes can be used to clip a signal, I tend to use a special 'clipper' plug in for this job.

Final comment, as always it is imperative that you don't forget to listen and to compare the un-mastered version with your mastered version. That means also that you loudness match between the two tracks. Obviously the mastered version will sound more compact, through the dynamic range minimisation after limiting, but pay attention to what the processing is doing. Has the tone shifted dramatically through the equalisations? Has the limiting or clipping introduced artefacts, or changed the transients in unpleasant ways? Does it feel balanced throughout? Does the master enhance what is there and bring it out in a positive way and address the issues you were trying to fix?

The final comment – use a mastering engineer for professional releases. Develop a relationship with a mastering engineer that you like and trust and work with them. The extra pair of ears, the heightened listening environment, the quality analog gear, the experience – is worth it.

Chapter 27

Impact of AI on mixing

In the year of this book's creation, artificial intelligence (AI) is taking hold and like most technology AI can be viewed negatively or positively depending on your perspective. It is important for me that AI be viewed critically like any other tool and assessed for its benefits to the job that it is trying to fulfil. One of the very important questions from my perspective is how much of the decision making power is using this tool taking away from me. At this point, we need to remember the opening chapters of this book when we defined what mixing is. Part of my definition of mixing involves making decisions – for me this is a central concern. If I give all the decisions over to an AI or multiple AI, then I stop actually doing the job of mixing, I have given authority over to the AI. Now for some people this is not a problem, but others this spells the end of their role in the process. We then need to dig deeper and get at different base level drivers for both people who mix and those who pay for it.

I would suggest that mixers broadly feel like they offer a creative development of the song from production and also feel that they have a certain style or approach as a mix engineer. Therefore this style is important to them. By handing over decisions to the AI, this will erode the style in favour of more generic or copycat approaches. By taking the own human interpretation out of it, this takes the creative control away from the human mixer. Of course I am talking here of extremes to make a point, but hopefully it is clear, and an awareness of this is important as a mixer choosing to integrate certain AI tasks into the mixing process.

Thinking about the question of AI and creativity from the perspective of those who pay for mixing will reveal different intentions that underpin peoples involvement in music. Those people who are more about money are more likely to be impressed by AI and chose to use AI and automated services over working with a human in order to save money. These types of people cannot be won over and it is best to not worry about it. Those who are very motivated by the creative pursuit will sit on the other side and see that involving AI too much in any process needs to be carefully mediated and critiqued in order that creativity is retained. There will be many people in the middle who want to craft an original sound but also want to save money where possible. These people are the ones that mix engineers will need to be in dialogue in the future.

Types of AI tools

At the time of writing this, the exponential growth of AI tools continues. In mixing, AI is not new, and it has been a part of many common tools that have been around for a long time. The current AI boom now combines with the exponential growth of computing power. What follows is a summary of some common types. It is not exhaustive but can help those of you new to AI to start to understand some of the common uses.

Adaptive processing for noise reduction: these types of AI tools are constantly analysing and adjusting in real time to the audio spectra. Think iZotope RX and other tools that are cleaning up audio.

Assistants: these types of AI tools will analyse and suggest approaches. These are very popular with mastering plugins, but also are finding their way into mixing. Many suggest EQ between different sounds that identify masking. There are also AI tools for dynamics.

Online full service for mixing/mastering: These services you upload your stems and the algorithm does the rest.

AI for emulation of analog processes: Rather than just applying an impulse response to the track, these AI powered emulations are designed to give a more authentic and adaptable response.

AI as a co-creative tool

AI certainly has some great benefits in mixing, in particular I find them useful for noise reduction and adaptive processing. They can speed up certain laborious processes definitely, but used unchecked, they start to create results that sound fine but often miss something that we can't always put our finger on. For the moment at least, AI works best as a co-creative tool where we can collaborate, we can analyse the suggestions and we can apply the changes that align with our own sense of aesthetic.

Future prediction

Mixing choices in the style of – AI that has analysed the approach of your favourite mix engineer or a specific song and can transfer this onto your mix. This is something that is not here yet, but I predict it will come in the future. This can be helpful as a learning tool certainly, but by relying on copying what is already there that is not distorted through the lens of human interpretation we will get more direct copies, which will just render music boring. I would already argue to an extent that the preoccupation with direct copies means that music is already becoming bland. There will be no doubt those who get on board with this, but as I said in the opening to this section – people who are attracted to this type of tool, are not interested in art or creativity.

Final word, for now...

Is there a future where AI mixes everything? Maybe, but there will always be a human who is listening and deciding if they like what the AI has done. AI is not interested in the artistic and communicative act of creating music, it is a human endeavour. It is part of the way that humans chose to spend time to figure out life, to share who they are and what they like and yes of course also to make money. But if we take money out of it for a second, humans started making music to share and feel connected to each other. Just because an AI can do it, doesn't mean that humans will stop doing it. The AI will take the jobs that give little satisfaction, but for each human that will be something different. You can find out for yourself what parts of mixing you would be happy to pass on. If you find out that you are happy to pass it all on, then guess what – you were never a mixer in the first place. If you are an artist who is happy to utilise AI for the entire mixing process, then that is also fine. As we will see increasingly mixing is becoming impossible to distinguish from production and as we can see that production is a key element of the aesthetic determination of a sonic piece, then if we give this over to AI, we are no longer being creative and no longer communicating our ideas, feelings, taste, etc.

Chapter 28

Spatial mixing – mixing beyond stereo

What is spatial mixing?

Spatial mixing, or 360° mixing or immersive audio are often terms used to describe a similar concept. This means mixing from the perspective of giving the listener an experience of sound coming from all directions. This means extending the field from in front to include behind as with the first quadraphonic and 5.1 mixes and now it has extended to include a sphere which includes height information as well. In the 1990s surround sound referred to mixing in 5.1, today we can achieve a spatial or immersive mix with a vast array of speaker set-ups, from 4 to up to 100+ speakers and, furthermore, there is also more than one way to approach it. Spatial mixing is largely a conceptual framework with many different ways to approach it as well as many different final outputs. It is little wonder that many people are confused! Never fear this chapter is diving in and going to consider spatial mixing from multiple angles, so that you can feel confident about considering the particular approach that suits you, your application and the desired result. This chapter is a starting point to help you get your head around the realm of spatial mixing. It is also recommended to dive in further for more specifics with various texts found in the expanded reading bibliography at the end of the book.

Is it a fad?

This is one of the most common comments that I hear so I feel it needs to be addressed before we dive in, often focussing on this question obfuscates the ability to investigate and learn about the technical and conceptual details. So many people have opinions that might not know so much about it! We can't deny that spatial mixing comes and goes. It is often touted as the next thing to replace stereo but then retreats into the sidelines. I can't predict if it will supersede stereo, though I doubt it. It is another concept that sits alongside stereo rather than replacing it. I can confirm that spatial audio has been enjoyed and invested in for as long as stereo has. The format is open and there are many approaches so this inability to standardise things gets in the way of it becoming a stable household concept.

At the time of writing this book, in 2024, there has been a significant investment in 'spatial' audio by Apple Music which has led to the rise of atmos mixing – that means utilising Dolby Atmos specific approach to mixing spatially. There is significant investment in this format with a desire to create a new standard. Whilst in some ways standardisation is helpful, in the case of immersive audio it seems to me that choosing a format is already part of the creative process like choosing what DAW to use. It is ok that there is no standard, but it does mean that it is harder for the mainstream to hold on to. Whether the push by Apple Music and Dolby Atmos is a fad remains to be seen, but the idea of immersive formats has been around for a very long time, and it will persist. Whether it is specialist binaural headphone listening, multi-speaker listening for private set-ups, immersive installations in art galleries, ambisonics for gaming or Dolby Atmos for cinema and headphone listening, immersive mixing is a skill that will remain an approach to being creative with sound and space.

As a passionate sound lover and maker I can honestly say I love stereo and I love spatial, but I don't love ALL of either. In particular, with regard the approach to space and immersion, trying to force all music into the new spatial format feels problematic for me. I do not feel that every sonic creation should exist in both formats. Spatial music is a wonderful experience, but it is rarely created by just remixing a stereo production into the Dolby environment.

The rise and fall and rise of spatial audio

The history of spatial audio is interesting because although it feels like a new phenomenon to many, when we trace the history we realise that it has been on people's minds since the invention of loudspeaker technology. We can see that improvements in mechanical, electrical engineering and latterly in computer DSP and digital technology lead to invention of new technologies which has created a cyclical resurgence of immersive concepts in the mind of the mainstream. We can also note that by following popular and experimental artist that pushed creatively and technically through their sound making we can trace the emergence of such new technologies. Artists like the Beatles in the 1960s and 1970s and Bjork in the 1990s and 2000s have jumped on these new technologies and explored them with gusto through their music. A consideration of the history of spatial audio is also important because we realise that there are multiple formats and approaches that have been developed in parallel. It is very easy to simplify it and think that immersion is only one thing. The loudest and most dominant brand might silence the varied innovations that run in parallel. So below we follow a brief and incomplete history of immersive audio (taken here to mean formats beyond stereo) starting with the invention of quadraphonic listening.

In the 1970s, quadraphonic mixing emerged as a new format for music making. There was significant investment and buy in from manufacturers. Large format mixing consoles that were built during that time had the option for front and back bus output for the speakers placed in front and behind. Four track and eight track tape machines were built that had the ability to output to multiple speakers, and special alterations were made to vinyl players in order that they could playback quadraphonic information. Buchla designed a synthesiser that was quadrophonic in nature. The stage was set for quadraphonic to take the world by storm and leave stereo music in the dust, however that was not to be. Quadraphonic came and fell back into the background for a few reasons one important one being that the average consumers could not afford or could not fit the heavy and cumbersome speakers into their home listening environments.

We saw the second wave of immersion take us by storm in the 1990s with the move to home cinema. Between Dolby and DTS it seems that cinema surround was finally to make its way into the homes of everyone. Although speakers now were much smaller, it was still expensive, and most people did not know how to set up their speakers at home, so this came and fell back into the background for home listening. The format has remained largely in commercial contexts where bespoke listening environments can be installed and maintained. We would be greatly disappointed if we went to the cinema to see the lates blockbuster and we were met with a simple stereo sound mix. So whilst home listening remained largely in stereo, cinema continued to develop the multichannel format, from 5.1 to 7.1, 12.1 and so on. These formats originally placed speakers around us at the same height, but the desire to hear sounds fly above us lead to the development of height positioning and into more complex formats such as 5.1.2, 12.4.2 and so on. These formats were generally dependent on the specific room size and the system was built specifically for those rooms. Highly specialised, highly expensive.

Quietly in the background the innovations originally presented by Blumelin and the concept of mid/side components in stereo were being manipulated to develop a new format called ambisonic developed by Gerzon in the 1970s. This had potential but lacked the resolution and clarity that was able to be achieved using channel-based methods. It was not until higher orders of ambisonic were introduced in the 1990s that the potential started to be realised. However, initially these potentialities remained only realised in academic and research lab context. It seemed at that time that spatial audio was destined to exist in commercial cinema context, for bespoke high fi obsessive with big wallets or in complex scientific technical contexts.

The situation remained as such until the last five to ten years. We are in the middle of immersions third coming. The rise of VR and XR meant that the topic of immersion began to bubble once more. In these contexts, the approach was predominately headphone based with binaural output favoured. Also during this time ambisonics recording technologies have been developed with many microphones hitting the market. A new area of music recorded with ambisonics microphones released as binaural developed. This included classical and field recording contexts where the enhanced realism of ambisonics microphones contributes to the genre.

Around 2014 the conversation about immersive audio in musical and sound contexts started to reemerge. Spatial audio that had found home to both cinema and experimental electro-acoustic composition and sound installation contexts started to re-enter the mainstream. This coincided with companies that develop software and plugins for VR such as Dear Reality. This meant that the technologies that were restricted to specialists and research contexts started to become more available in the consumer market. We started to see special performances that were presented spatially, the ground swell of change had begun. In 2021, a major shift occurred when Dolby Atmos was made available for consumer consumption. Consumer music production software Logic Pro became compliant with atmos format without requiring additional hardware that had previously been the case. Dolby Atmos had been introduced in cinemas in 2012 and the integration had existed with Pro Tools users for some time as Pro Tools is the traditional choice for surround mixing for film. What was new was the integration of Dolby Atmos into Logic Pro and additional tools that meant the Dolby renderer and panner encoder could be connected theoretically to any DAW. This was also supported by Apple Music who together with Beats integrated the binaural spatial movement with their headphones. Spatial audio was now definitely in the mainstream once more.

The last couple of years we have seen a massive push for music to adopt this new approach. Significant money has been spent not only mixing new albums in Dolby Atmos but also going back into old sessions from many of our favourite albums and re-releasing

158 Mixing in Flow

them in atmos. In 2024, we can see that the crest of this frenzy has past and things seem to be settling down somewhat. Although perhaps at this particular moment it is too early to predict, it seems to me that the spatial format will stay around but the mad obsession will fade. This means that it is not likely to be adopted into the mainstream and replace stereo, but it will sit beside stereo mixing. I have some more opinions and predictions on this that I'll also take up in more detail a bit later, but for now let's get into the details.

Speaker layout

I am starting the discussion at the end of the chain – the monitoring – because this is an important concept to keep in the mind as we move from stereo into more complex playback formats. Let's go through various ways to monitor audio and discuss their uses and traditions.

One Channel: We start with one speaker, mono, which was the way one monitored recorded sound in the 1920s with the first phonographs. Even though this is very old school, you may be surprised to know that mono listening still occurs in some instances. For example, sometimes radio broadcasts as mono, because of the bandwidth required to transmit two signals. Also, in some dance club formats, the speakers are set up mono, to ensure that the stability of the signal is maintained throughout the club independent on where you are in the room.

Two Channel: most popular format that has been the preoccupation with the book until now – stereo format. This is characterised as two speakers equidistant from the listening position and each other. Also in the two channel format is binaural. Although not exclusively, this is often optimised for headphone listening.

Four Channel: Moving along we then find ourselves investigating the field in front and behind with the introduction of two more speakers in the back. This is called **quadraphonic**. There are two basic approaches to the placement. (See Figure 28.1).

Six Channel: Next is the 1990s arrival of **5.1**. This was first introduced in cinema and then presented for home cinema. It expands on quadraphonic by adding a centre speaker (for dialogue) and also a sub speaker for the ultra-low parts of sound (for impact and drama). This format stayed for quite some time with cinema adding more speakers to create basically a sharper resolution and 'life-like' experience for the listener. This started with more speakers on the side 7.1, 12.1, etc.

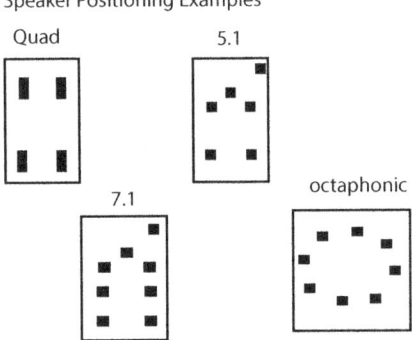

Figure 28.1 Different approaches to speaker placement

Eight Channel: In parallel to the home cinema approach, we also have different approaches being favoured by experimental music contexts, electro-acoustic and sound installations. A circular installation of speakers can be **octaphonic**. We can also go to the extreme and consider a wave field synthesis system which utilises a large amount of speakers tightly aligned next to each other.

14 Channel: In 2012, we see the emergence of Dolby Atmos, which utilises the existing Dolby format 7.1 (and variations) but extends the basic configuration to include height speakers, thus making it a true 360 format as well. Dolby Atmos is interesting to pause the discussion here, because Dolby has done the most to try to establish a new industry standard, with standardisation of the speaker layouts **to 7.2.4**.

20+ channels: Loudspeaker domes in bespoke venues. Wave field synthesis.

We can see from this opening section there are so many ways to explore 360/immersive sound. With all these different speaker set-ups, there can be different benefits and drawbacks. It can seem overwhelming, but the good news is for beginners in particular, with the invention of object-based mixing we can largely ignore the above details about the final output format when we are mixing and focus on creating a quality immersive experience in whatever speaker configuration we have access to (that includes headphones). More on that below.

Different approaches to spatial mixing

There are several different ways to approach spatial mixing. Clarifying the difference between each of these can be difficult. Each person has their own approach that they favour. Below we go through the key concepts to help you establish an overview and the ability to start to consider what might suit you. It will include basics about the set-up, including physical requirements and discussion about the approach to panning as well as discussion about potential limitations.

Channel-based multichannel (e.g. 5.1)

Listening Examples: Beatles, 5.1 Film score, Susanne ciani quad

Channel-based audio is what we are all used to. Stereo mixing is an example of channel based. We mix to a master output that has two channels of audio which are sent to our two speakers left/right. There are many different types of multi-speaker set-ups that we might utilise but the main thing that needs to be understood about channel-based mixing is that you **must use physical speakers**. For example, if you want to mix in 7.1 – you must have seven speakers and a subwoofer. These speakers also need to be set up correctly and tuned properly.

The surround panner

We use amplitude panning that is most likely based on constant power or compromised/compensated approach. (For more information, look up pan laws). This adjusts the level going to each speaker based on the position of the pan pot (physical/virtual). In channel-based mixing, the normal stereo panner gets a boost and becomes the 'surround panner'

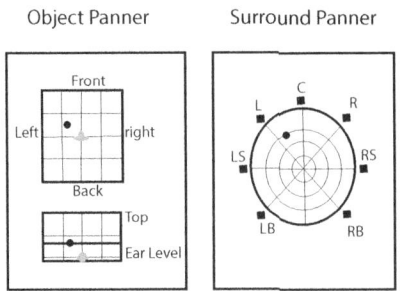

Figure 28.2 Surround panner example

which operates on similar principles to the stereo panner but includes more speakers. The surround panner will offer a visualisation of the physical set-up of the speakers in the room you will position the sound where you want it to sit within this physical arrangement. The surround panner (Figure 28.2) will often also have a SUB or LFE option where you can choose to route the signal off to the sub speaker, sometimes also known as the LFE channel.

Limitations

We are stuck with mixing for the actual speakers that we have in our room. The panner is generally limited to the basic channel base formats (quad, 5.1, 7.1)

What do I need to do a multichannel mix?

In order to mix in multichannel formats, you need at a minimum the following:

- An audio interface with 4 or more outputs
- A DAW that allows for multiple outputs and multiple busses (Logic Pro, Cubase, Reaper, Pro Tools), or if not multiple busses, then it must offer a routing solution (as does Ableton)
- Multichannel signal processing (plugins)

Note: This is why Ableton is not usually the choice for people mixing multi channel – Ableton only allows stereo busses, which also means that it cannot apply multichannel plugins. There are workarounds, but with other DAW providing easier options, for most it is not worth the hassle.

Object based

Listening Examples: Electro-acoustic and experimental artists, Kaija Saariaho

Object based takes a rather different approach to positioning sounds in space. This concept maps a sound in a virtual three dimensional field, using the Cartesian coordinate (xyz). This information is stored as metadata. The original audio source file (mono audio stem) is stored with this metadata and the positioning in space is rendered in real time by a computer program or a devoted hardware renderer.

Because this positioning is rendered in real time during playback this means that we can easily change the output configuration. That means we can monitor with different speaker configurations. You might work at home in headphones one day and set the output renderer to headphone(binaural), the next day you might be in your studio which has a 7.1 set-up. The positioning is stored as metadata that can easily be mapped to each speaker configuration.

The object panner

Rather than a surround panner that echoes the physical layout of speaker configuration this is a panner that represents the virtual 3D field where the sound can be 'mapped to'. Positioning this panner records metadata that is stored with the sound file.

Limitations

This is a newer format that is not included in DAW's, it often requires a greater learning curve to install or a significant investment in third-party software, but this will likely change, as in my opinion it is a great option that will increase in popularity in the coming years.

What do I need to mix in object-based formats?

- A DAW that supports this – Ableton has extensions through max4live
  OR a third-party program that acts as renderer, e.g. Spat Revolution or L-isa.
- Reverb or spatial addons to support sounds moving naturally in 3D spaces rather than in 2 dimensions of stereo processing (most software include these in the package)

Dolby Atmos

Listening Examples: Apple Music, Cinema

Dolby Atmos combines both the channel-based concept and the object concept. It has evolved through the cinema format and recently found its way into music through the collaboration with Apple Music. Because of the history in cinema some of the terminology reflects this For example the channel components are called Beds, referring to the way that film sound traditionally categorises different sound layers. These layers use the surround panner and work as normal in application in a mix. The difference comes in that the bed subgroup (multichannel file for example 5.1) is rendered into the 3D object space. It is a bit confusing to imagine before you have interacted with this format, but the important thing to understand here is that both the BEDS and the objects are rendered into the virtual field which means that the output listening environment is not fixed like channel based and behaves as object based.

Two panners! More routing

Because of this combination of two approaches Dolby Atmos can be a little more complex to set up. It involves deciding on which elements you want to exist as beds and which you want to be objects and then you must rout them accordingly as well as use the correct panner (Figure 28.3).

Limitations

Probably due to the history in cinema and the continued use of channel-based beds, the speaker layout options available in Dolby Atmos renderer are limited to those from cinema. Thus, those wanting to work in more experimental, installation or dome configurations are not able to render to these speaker formats.

What do I need to mix Dolby Atmos?

Dolby Atmos comes bundled together, so you get the renderer, panner and other essentials when you work with either the stand alone Dolby Renderer program or if you have a DAW that supports it.

The important thing to note is whether you have a DAW that includes Dolby Atmos built in, or whether you are able to link to the external renderer in your DAW.

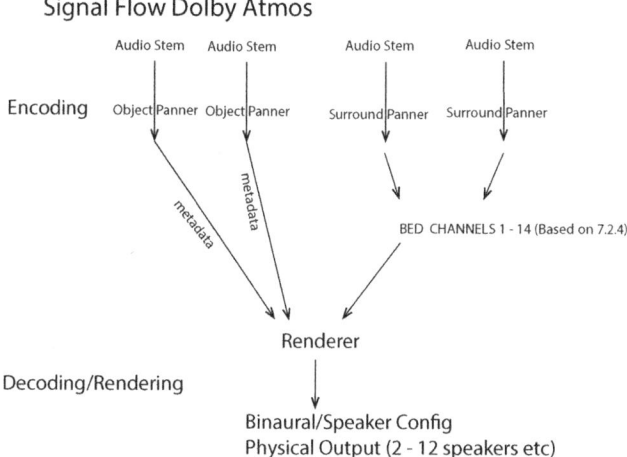

Figure 28.3 Routing in Dolby Atmos

Ambisonic

Listening Examples: Natasha Barrett

Ambisonic appears on the surface to be very similar to the object concept. Similar to object formats we are able to change the output configuration because we encode the speaker positioning into a field. Different from object based however, we do not create any metadata rather the positioning is encoded into a **sound field.** In the most basic description, we understand a stereo miking technique to create a basic sound field. We can extend that principle to consider multiple stereo miking techniques that extend the field to include not just in front but also behind and above and below the sound source. Therefore, capturing a surround sound field. Another important point about ambisonics is that because we are using the sound field to encode the position information we utilise multiple channels of audio. The multiple channels become the resolution of the encoding. So a simple 1st-order resolution is four channels whilst the most complex is 7th order which utlises 64 channels of audio. Ambisonics is a very open format there is a great deal of flexibility and customisation with this approach.

Routing

Routing is an important component of ambisonic to understand. The number of channels used in the busses relates to the resolution of the sound field. We must ensure that the routing is set up correctly for the sound field to be correctly encoded (Figure 28.4)

The encoder/panner

Similar to the object panner this panner represents the 3D field and we place our sound within it, but more like our experience of stereo panning this positioning is encoded into the sound field which is routed out a multichannel bus size depends on the resolution of the project.

164 Mixing in Flow

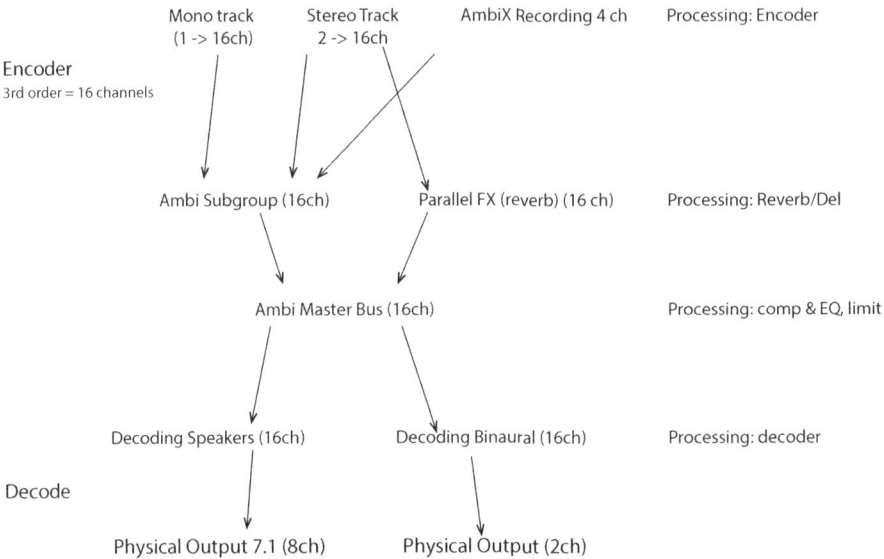

Figure 28.4 Routing in ambisonic

Limitations

Because it uses sound field or phase relationships to encode physical positioning, ambisonics traditionally features a specific sweet spot, meaning that if you want to walk around the sounds will change as they move in and out of phase.

What do I need to work in ambisonic formats?

- Reaper
  In the past, you would be able to work in Pro Tools and logic, but with the development of Dolby Atmos this has all but been abandoned in these DAW's. The most common solution is to work with Reaper. Reaper is the best solution because it allows up to 64 channel busses which is the highest resolution required for ambisonics routing.
- Third-party ambisonics software – for encoders decoders and multichannel processing. There are many available for free: IEM, Sparta

Ambisonic has been developing in academic institutions and as such many of the tools available to work with it are free or very cheap to purchase.

Advice for choosing which approach to adopt

When considering which approach is best to start with, at first, you might want to pick the easiest to grasp. In this case, it would be channel based. This is an extension of stereo, and you have the physicality of the speakers that is replicated in your DAW with the panner. The problem with this is that many people starting out might not have the space or the availability of adding more speakers or using an audio interface with 6 outputs.

If you will work in headphones for at least the majority of your work (although testing with physical speakers is highly recommended), you will have to pick either object (which includes atmos) or ambisonics.

Many people are confused between ambisonics and object-based approaches. Whilst they do share some similarities there are some big differences. Mainly in the workflow but also in the ability to customise. Computer and DAW compatibility might lead you to one approach or another, as might budget constraints. Wherever you stand just pick the approach that fits you best and offers the easiest entry into this new field. Once you pick an approach explore it thoroughly, create some mixes before moving over to another one this will avoid confusion and ensure that you have gone deep enough to establish your preferred approach.

If you want to dive in and get professional from the start, then the two dominant professional solutions are Reaper for Ambisonic and Pro Tools for Dolby Atmos.

Depending on where you see your future as a mixer, if it is in the world of games, experimental and such, then Reaper is a great solution, however if you see yourself mixing for film and mixing music for the new spatial apple format then Pro Tools is really the DAW you need to be using.

Key concepts of spatial mixing

Regardless of whether you mix with channel based, object based, Dolby Atmos or ambisonic, there are concepts that are shared across all formats.

Monitoring

This refers to how you are listening to the mix when you are mixing. If you are doing a channel-based mix, you must monitor the same configuration that you are exporting. So if you are doing a 5.1 mix you need to have 6 speakers. Most often though these days with Atmos, Object and ambisonic formats most people enjoy the fact that the way we monitor whilst mixing is no longer tied to the output format. If you have headphones, the monitoring format would be binaural. You might have the physical speakers in your mixing room such as 5.1 or the Dolby standard of 7.1.4. Perhaps you are doing adjustments in a more complex configuration such as a Dome. All of these are different monitoring formats. Remembering that, unless you are doing a channel-based mix, you can change your monitoring format and render out mixes in different configurations than what you are monitoring. For example, if you are monitoring a Dolby Atmos mix with headphones, you can render to 7.1.4 or even leave it as a Dolby BWF file.

Binaural mixing/monitoring (headphones)

This is when you monitor/render/decode your 360 spatial environment to two channels. As you can imagine there is quite a great deal of complex processing going on under the hood. For this, you will be using HRTF (Head Related Transfer Functions). These are calculated based on the shape of the human head, taking into account how this impacts sound reception. More on this below. It is important if you are working binaurally to take a moment to understand this concept. After reading below you will then need to go online and search for files and get ready to do some comparative listening to find the files that give you the best spatialisation. Give yourself time to explore these files with your ears, comparing and contrasting until you find a good match that gives you a good comparison in binaural verse spatialised. Read below for more information on HRTF and SOFA files.

HRTF (head related transfer functions)

The shape of our head, ears and also their body effects the way that we perceive sound. When we are working spatially having speakers positioned all around us, this phenomenon becomes more pronounced. The development of binaural mixing whereby an ambisonics microphone has recorded a sound field or perhaps in a DAW production a producer has crafted a 3D sound mix (using one or other of the techniques discussed above) the headphone listening experience must consider these differences in order to create a realistic three dimensional rendering. Many of us have used 'spatial' headphones, if any of you have the Apple headphones you will have undergone a set-up process where the phone takes a picture of your head and ears. This process was creating a specific analysis of your physical characteristics and matching it with a file that represents your head best to ensure the optimal HRTF are applied when listening.

SOFA files

These files contain HRTF that have been captured and calculated from a range of different heads. There are many databases online where you can find an array of these files to download and try out. Whilst there is no perfect way to go through the files, my method is to search through the databases and try to find a head that resembled mine, then I downloaded a selection and tried them out.

Once you have the SOFA files downloaded, you will need to upload them into a binaural decoding plugin that will read them. One of my favourite binaural decoders is the Ambi HD Decoder by NoiseMakers. The better match you can make to your own body the more accurate the binaural rendering of the 3D listening will be.

Panning

Panning in spatial contexts can be understood relative to stereo, but it becomes much more complex because there are so many more fields to consider. If we think of the most simple extension channel-based we have five speakers now instead of two, the panner turns into a surround panner and we can move the sound in a circle in 360°. When we start to move away from planar or periphonic (meaning all on one plane) adding height speakers, we can now move sound in more directions that left and right and forward and back. It gets even more detailed when we introduce the 3D object panners and we are able to map the positioning of an object space.

If you are working with an ambisonic or object panner, you will also find that the panner acts as the encoder which converts the mono/stereo file into the 360° sound field. If you are working with Dolby Atmos, the object panner records the positioning of the object in space and renders that information as metadata which is stored with the raw sound file and then every time the piece is played back that metadata maps the positioning of the objects in space and renders it out to the speakers in real time.

Panning becomes a real art in spatial contexts and as such a great time an attention can and should be placed on considerations around panning. The tendency for beginners is to make things spin around, and whilst that is fun at first, there is more success in paying attention to the detail in crafting an impressive 360 experience.

Spatialising

This term gets thrown around a lot in relation to spatial mixing, but really it is just an extension of panning. We are still technically spatialising sound in stereo mixing, but we don't really refer to it in that way. Spatialising is also used to describe the process of automating the movement of an object (sound) in the 360° space.

Encoding/decoding

Encoding is the process of converting or combining a sound file with its positioning in 360 space. This is quite confusing to many people when they first encounter ambisonic mixing. In ambisonic mixing, a 360° sound field is created when you position your sound within the 360° field. This becomes simpler to understand if we consider that an interleaved stereo file that we all work with also has positioning information encoded into the file. Just now with ambisonics there are many more channels representing the level of resolution of the 360 sound field, so the concept seems more complicated.

It is very important to understand that once a sound is encoded it must be monitored in the correct way in order that the spatial information be properly maintained. Audio is still heard but the sound field will be incorrect. First time ambisonic mixers often make mistakes, understandably it is confusing. Good workflow that includes proper set-up and checks along the way will ensure this doesn't happen to you.

Decoding is what happens at the end of the chain in order that we can monitor it whilst working as well as render out the final project when we are finished. The renderer will contain information about the speaker system you are using to monitor the project. You might use a binaural renderer as discussed above, or you might have a specific speaker configuration that you are decoding to. Dolby Atmos uses a slightly different name for this stage, they call it render. More on that below.

If you work with channel-based formats, you will not have to consider encoding/decoding.

Rendering

Monitoring

In Dolby Atmos, they call the whole process of monitoring rendering. So the concept above discussed in relation to decoding, is referred to as *rendering* when working in Dolby Atmos.

Bouncing

Rendering is the final process of the spatial mixing process This is where you decide on the final material output for the work. You might be monitoring on one way, but want to render it a different way. No problem. It is important to understand that monitoring and rendering are two related but different concepts.

If you want to finally output your project for a **fixed speaker set-up**, for example 5.1 is still quite common, you can do this. You can also choose to leave the file in the 360 format so that the final speaker arrangement can be selected (rendered) at a later time and place. If you want to leave the final speaker configuration open when working with ambisonic files, you would render them in a format called **ambiX** the resolution is determined by which order/how many channels you will render out. For Dolby Atmos, you bounce a **BWF file** that flattens the BEDs and leaves the objects and metadata together as separate audio tracks.

If you work with channel-based formats, you will not have to so much about rendering because you monitor what you render – you bounce in a similar way that you do when working in stereo formats you just need to ensure that you have decided if you want to render individual files or interleave the mix.

Routing/bussing

Routing is a very important and sometimes tricky part of working with spatial formats.

In all formats, it is important that you are aware of where and how you are sending your sounds into the 3D field. The combination of routing and panning forms substantial amount of the creative approach to spatialisation in spatial mixing. A great deal of fun can be had. Review the specifics of routing for the format you are working with above.

Special reminder: In ambisonics formats, the amount of channels and the size of the buses directly correlates with the resolution of the overall system. A great deal of care needs to be employed to the routing when setting up your ambisonics project. For this reason, templates are often a great solution.

LFE

This refers to Low Frequency Environment and refers to a channel that is reserved for low frequency or sub, the sound is sent to a special speaker to handle low frequencies. Beyond separating the low frequencies in the mixing project there can be special set-ups in monitoring which relate to whether bass management is set up or not. Bass management means that the satellite speakers have a low-cut so that all low frequencies go to the sub. This should be set up and tuned correctly by a professional acoustician (you can also learn how to do this yourself). If the system has bass management, it is important to be aware that low frequencies from satellite speakers are cut. Therefore it is important when mixing to pay special attention to this, particularly if you don't know what the systems will be in future playback environments. Conversely without bass management there can be increased issues with monitoring consistencies due to overlap between the speakers. There may be phase issues, as well as modal problems. Regardless of the monitoring approach we must remember as always that low frequencies need special attention and care.

Best practice tips

Moving from stereo to spatial

When moving from stereo to spatial mixing, there is an important shift that must occur conceptually as well as practically. Key mixing processes terms (dynamics, frequency, balance, envelope, etc.) are still valid but often there is a new perspective as we expand the environment. For example, our approach to frequency and equalisation remains similar on one end, we still juggle individual sound relative to the whole and shape the frequency accordingly, however we are also considering upper frequency to enhance localisation as well as impact through use of sub frequencies.

In addition to concepts that are similar but altered, there are quite a few new concepts that you might not have encountered before. You will be hearing new terms like encoding, decoding, rendering, object panning. The set-up in spatial contexts often takes longer and can be confusing at first. In particular, the routing is an area that needs a moment of adjustment. Taking a moment to fully grasp the new concepts as well as taking time to understand the reasons behind the different set-up and workflow will lead to a smooth mix. Following is a broken down guide sharing key elements to incorporate into your approach to spatial mixing that can help you find your way through this new process.

Prep the audio

It's a good idea when you perform your first spatial mix to start with audio stems where the production is already well developed so you can focus on the mix and spatialising.

The first tip is to start with the audio in good shape – that means the sound design and arrangement should be completed. It is also a good idea that you have the production at a fairly solid pre-mix level. Note: this does not mean they should be over compressed with dynamics removed. In spatial mixing, it is important that the dynamic quality of sounds is maintained, over compressed stems can yield an unsuccessful spatial mix.

Pay attention to mono/stereo. Where possible turn your stereo sounds into mono. Only keep stereo sounds when you are absolutely positive that you want the stereo information. Remember that stereo sounds create a diffusion so that the source localisation is not as defined. This might be something that you want which is fine, but just make your choices. It makes things easier down the road.

Important note: It is necessary to work in **48k sample rate** (or 96). For Dolby Atmos – the renderer will not work properly. For ambisonic, some of the better plugins will not work (SPARTA Decoder). It is best practice to just get used to working in 48k for all spatial applications.

Setting up and routing

It is very important to take time to set up the routing correctly. You might use a template that already has some of this formatting set-up, or you might develop your own approach and build your own template. Whatever you do there will be some manual work involved, **it is imperative that you test this set-up before you start mixing**. Every time I skip the testing it always ends up that something was incorrect which gives me unexpected results slowing me down as I troubleshoot to find the problem. In ambisonics formats, you need to decide on your resolution and set up your channel and bus outputs to that number. Third order ambisonics is quite common which uses 16 channels but if you are working in Reaper you can extend that to the 7th order that uses 64 channels. When working in Dolby Atmos, you have the important task of deciding to send your sounds to the beds or whether to create an object. Once you are working with the object its output must be correctly set up and received by the renderer in order that the spatial information is recorded.

The key processes of mixing are still frequency and dynamics

As already discussed, just as in stereo mixing, equalisation and compression are still a key component of spatial mixing. This doesn't mean necessarily mixing from scratch. As we have changed the spatial relationship as there will likely be further adjustments necessary to get the balances correct as well as to enhance the position and localisation. One of the main EQ processes that I often perform is low mid reduction, perhaps due to less masking than occurs in stereo contexts. With regards dynamic processing it often relates to envelope and ensuring that the transient information is optimised for localisation and working with other tracks spaced in other areas.

IMPORTANT NOTE: These are often performed before the encoding. After encoding you must use specialist plugins that work on the encoded signal.

Low frequency management

It is vital that you are purposeful with low frequency when mixing spatially. Awareness of low frequency is hopefully not something that is new to you as we always consider low end when mixing in stereo formats. However, the approach to dealing with low frequency is different in spatial mixing because we have a dedicated channel and speaker for these sounds.

1. Decide what sound you want to send to the low frequency channel – then send them
2. Consider the loudness in relation to the other instruments
3. Consider the tuning of the low frequency that is being sent
4. Don't assume that there is an automatic cut off applied to the satellite speakers – maintain clarity by clearing away the low frequency information for sound that you are not wanting to hear the low frequency information

Note on Ambisonic and SUB: It is actually more critical in ambisonic mixing as opposed to channel based and atmos, because in ambisonic the low frequency is calculated and sent to the sub – so if you haven't cleaned up the low frequency it can create problems. The sub-information is the channel 1, the mono or the 0 order in the ambisonics channel routing system. This is folded into the channel matrix. In the traditional channel-based format, you would create an LFE bus and this made it very clear what information is being sent there.

Loudness

One of the reasons that I love spatial mixing is that the 'loudness wars' debate becomes redundant. With so many speakers you don't have to worry about loudness. Moreso, if you push the limiter too much it really spoils the mix because the consequence of loudness is reduced dynamics and dynamics are important for depth perception which is a vital aspect of realising spatial mixes. I have heard spatial mixes by first time immersive mixers that were mixed according to the stereo loudness approach and it is basically unlistenable and definitely not spacious. Maintaining dynamics is important in these formats. This is why it is important that the stems you bring into the mix are not already over compressed.

Depth and space

One of the key aspects of creating an engaging mix is an awareness and crafting of space. This is the same for stereo as it is for spatial mixes. The mechanics however of working with multichannel delay and reverb plugins are not simple. It requires many times the processing, and when working with convolution reverbs requires that ambisonics impulse responses are being used. These specialised reverbs are often very expensive. If you work in Logic and Dolby Atmos, you can use the space designer reverb as reverb that sends to beds. Ambisonic reverbs such as NoiseMakers ambiVerb is great it is limited to the 3rd-order resolution however. Working with the free IEM plugins fdnreverb allows up to the 7th order (so that is a 64 channel reverb!), it is a solid reverb but it doesn't dazzle with realism as some other impulse respond verbs do however. Commercial reverbs such as Exponential Audio's Stratus and Symphonie offer exciting options that can work across multiple formats.

Multichannel bussing and parallel processing of reverbs is an important aspect in my opinion when working in spatial formats. For this reason, I do not recommend working with Ableton generally speaking for spatial mixing. There certainly are new formats such as Envelope4Live and bespoke max4live patches from immersive venues such as Monom in Berlin. The new Spat for Ableton packages that have recently been released are the first package that has created a streamlined, stable, great sounding and importantly clear 360 sense of space and localisation.

Multimono reverbs are a solution that are used when a multichannel reverb is not available, this can work but the detail and the true impact of sounds moving and reflecting through a multichannel space make the mixes generally not as convincing.

NOTE: Remembering now we can still use stereo reverbs to create specific spaces in different fields or planes within the 360 environment. This can be a really create aspect of working with space in immersive contexts.

Spatialisation – placement

Source positioning in spatial mixing are perhaps the most important decisions that we make in spatial mixing. Similar to stereo mixing, considering the balances of the elements and establishing of their positioning largely tune through panning and fader level. In an ambisonics, surround or object mixing, the panning is far more complex than a stereo pan and becomes a primary tool in the spatial mix. The mix must have stability and foundation just as any mix, so the first step is often to decide about the elements that will stay stable, which of those will be centred as well whether the centring is front or is above or below. Next we consider those elements that will be fixed across the 360 field and where to place them, considering the relationship to each other and the balance of the overall listening experience.

Mono or stereo sources?

Considering whether to import a source as stereo or mono is another very important decision. The current trend in popular music is to deliver stereo stems for all tracks, even kick drums. Exporting mono is a separate and time-consuming process and most don't bother. However, when in the immersive mix, it is very important to mono out mono sources first of all starting make your decisions about what aspects you want to consider

Mono vs Stereo Localisation/Diffusion

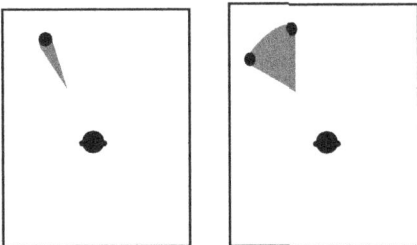

Figure 28.5 Mono vs stereo localisation diffusion

mono and that you want to consider stereo. When using stereo sources, or a stereo sub-mix bus, these stereo sources can create phantom image which may create phase cancellations as the listener walks around the room. In addition, when using a stereo source it creates a field rather than a specific object, this means that localisation will not be as clear or direct. So if we want a sound to exist in an area or a field then stereo can work well, but when you want to automate a sound's movement you will generally heighten the perception of the object spatialisation when working with mono (Figure 28.5).

Spatialisation – movement vs stability

Movement is an exciting feature in spatial mixing. However, is also something that is overused and often misused. At first, it is exciting, but this excitement is short lasting if it is too much. If the movement has no concept or reason, it quickly becomes uninteresting and works against the overall momentum of the piece. It pays to think very carefully before creating movements. Not all pieces require movement. Most importantly when you are automating elements ensure that you have already set up stable components that will enhance the movement through their stability. Now with that said, we can still acknowledge that there are great uses for it when designed carefully. Automating movement and making interesting shapes can create a lovely dynamism and life to the music. Using LFO's and envelope followers can create interesting textures and relationships between elements. Often it can work quite well to consider spatial automation in relation to gesture and the creation of spatialised shapes or structures, this also helps to ensure you are not just moving something around for the fun of it but have consciously crafted sonic sculptural objects in your mix.

Workflow

A recurring theme in this book is the importance of workflow. The emergence of consistent workflow represents mastery and confidence. Below is an example of a general flow through a spatial mix. This is a guide, and it is something that you can compare with your own tendencies once you develop your own consistent style, then throw away my guidelines!

1 Open up a template for the format I'm working with (this will have key tracks, plugs and routing set-up)

2. Import audio stems from the stereo mix export, then individually place the stems into the template tracks or duplicate template tracks
3. Consider the groupings of the stems – i.e. Drums, Perc
4. Set up encoders on individual imported stems
5. Once all elements are encoded then check the routing to find any anomalies before going any further
6. Consider the spatial placement of individual items – this is a BIG part of the spatial mixing that shouldn't be underemphasised. Simple in action but important for laying out the space and starting to get meaningful interactions across space
7. At this point, quite often I realise there are some elements that might need further adjustments to source character through frequency and dynamic profile adjustment
8. Consider the spatial planning and set up motion/gesture characteristics on specific parts (which means automations around the space)
9. Consider the use of reverberant spaces – set up different rooms to further create a solid sense of space and spatial placement – combination with using stereo reverbs and delays to create specific experiences of sounds
10. Address the bass management and ensure that only sounds with sub intention have low frequencies
11. Consider the intention of the music or sound piece piece and pay attention to details – transitions lead moments and ensure that these are not sacrificed due to the spatial approach
12. Apply master bus processing and finalise the mix
13. Render: being careful of the Routing settings and channels, often it is good to render multiple format, such as binaural, channel and the master format (e.g. ambiX, Atmos)
14. Check the renders

I hope that this chapter has revealed the openness of working with this format. It can be overwhelming at first, but if you pick one approach first and have fun exploring that for a whilst before moving on to another approach then it is much easier to handle. I have presented the information about the various approaches to spatial mixing in a largely integrated fashion on purpose to highlight the similarities between them rather than see them as greatly opposing forces. My style is to work with all approaches and be flexible to move between them as the industry trends and specific project dictate. If the trend of its evolution continues, we know that it will continue twisting and turning, so to be a valuable spatial mixer and producer the flexibility to work across different formats will be beneficial as well as it enhances a deeper understanding of what makes a sound immersive or spatial to begin with. To begin with however, take it step by step and focus on one approach at a time. It is a wonderful and open creative space to explore the fascinating relationship between time, space through sound.

Final word

Rules are meant to be broken

We have come a long way in this book. We started out with simple definitions and prescriptive recommendations, however as the book developed these direction became looser encouraging more openness and individual decision making. You might consider my approach at times a bit schizophrenic, at one time telling you do it this way and then next saying do what you want. This is by design. When someone first encounters mixing it can be overwhelming, we often find ourselves overloaded with choices a seeming avalanche of new tools to learn and great masterpieces to analyse. In the beginning, following rules is a way to block out some of the noise and focus on building core skills and gaining a foundational knowledge. At the same time, I have been fostering your critical engagement with the process developing your own response and allowing you to make decision based on your specific situation and your style and most importantly to respond to the sounds that are in front of you.

Break my rules, but always listen to your instincts

It is also a good moment to recall the section in Part Two of this book on embodiment and how learning to listen with our body helps us pay attention to sound in a different way. Listening with our body also helps us to tap into our intuition and our unique energy field. Remembering also the discussion about building your own style, we now can say – go for it, trust your intuition, your skills and your training and be confident to make choices based on what you like, what gets you excited and what you feel is right. This is a highly technical stage of sound production, but it is after all still creative. We follow the masters to guide our path until we start to forge our own. The rules are meant to be broken. So go for it. Have fun and create music and sound that is meaningful to you!

The learning never ends

The goal of this book is to help you cultivate an approach that fosters moments of flow – those times when you are fully focused, enjoying yourself and able to tap into your creative problem-solving skills. To enhance your ability to achieve this state, it's essential to prioritise the development of key skills. These skills are categorised into four facets of mixing: critical listening, technical processing, analysis and application.

While anyone can mix, few can do it well. By becoming more critical and specific in your approach and choices, you can join the ranks of those who excel in mixing. Instead of providing a one-size-fits-all method, I encourage you to question everything and seek the answers within yourself. Your unique responses to sound are valuable, and cultivating your analytical and decision-making skills will boost your confidence in your personal style. As we navigate a world increasingly influenced by AI, the focus should not be on prescriptive approaches. Instead, we need to foster critical engagement and aesthetic development, ensuring that this craft thrives well into the future.

I hope you keep this book by your side as you embark on your mixing journey. It's designed to help you interrogate your process and hold you accountable, ensuring that your work stems from a place of knowledge rather than mere mimicry. Once you start mixing, keep this book in your studio library. It's a valuable resource to return to throughout your life as a mixer. Mixing can sometimes feel isolating, and there may be times when you seek advice or a second opinion. While having a mentor is invaluable, let this book serve as your support as you navigate challenges, overcome creative blocks or seek inspiration. I hope it becomes a source of knowledge and encouragement – a trusted companion on your solo journey in mixing.

Recently, I had the privilege of meeting two of my mixing idols: Bob Power and Russell Elevado, the masterminds behind the iconic neo-soul sounds of the late 1990s and early 2000s. As a young aspiring artist, I was captivated by seminal records like D'Angelo's *Voodoo* and Me'shell Ndegeocello's *Peace Beyond Passion*. In my quest to understand how these masterpieces were created, I discovered the role of the mixing engineer, which felt like uncovering the key to a puzzle I had long struggled to solve. This revelation has since opened countless doors of sonic exploration in my life, deepening my passion for sound and music and allowing me to gain new perspectives on working with audio.

I share this story as a parting thought to emphasise the joy of being an eternal student of your craft. Each day, whether I'm teaching mixing, performing a mix or analysing music, I learn something new about sound and my approach to it. I've come to embrace this infinite journey, finding fulfilment in the process rather than chasing an imaginary endpoint of mastery. Remember, we all have teachers, and those teachers have teachers, creating a vibrant community of individuals passionate about sound – one that you are now part of. Embrace the opportunity to learn from others, wear your influences proudly and don't hesitate to share your unique perspective and style.

Bibliography and expanded reading list

Cox, Christoph, and Daniel Warner, eds. *Audio Culture: Readings in Modern Music*. New York: Continuum, 2004.

Csikszentmihalyi, Mihaly. *Flow: The Psychology of Optimal Experience*. Nachdr. Harper Perennial Modern Classics. New York: Harper [and] Row, 2009.

Everest, F. Alton. *The Master Handbook of Acoustics*. 1st ed. Blue Ridge Summit, PA: TAB Books, 1981.

Farnell, Andy. *Designing Sound*. Cambridge, Mass: MIT Press, 2010.

Howes, David. 'Embodiment and the Senses'. In *The Routledge Companion to Sound Studies*, 24–34. Abingdon: Routledge, 2019. https://search.ebscohost.com/login.aspx?direct=true&db=ram&AN=A1969624&site=ehost-live.

Massy, Sylvia, and Chris Johnson. *Recording Unhinged Creative and Unconventional Music Recording Techniques*. Milwaukee, WI: Hal Leonard Books, 2016.

Merleau-Ponty, Maurice. *Phenomenology of Perception*. International Library of Philosophy and Scientific Method. London: Routledge, 1962.

Mixerman. *Zen and the Art of Mixing*. Milwaukee, WI: Hal Leonard, 2010.

Nancy, Jean-Luc, and Charlotte Mandell. *Listening*. 1st ed. New York: Fordham University Press, 2007.

Owsinski, Bobby. *The Mixing Engineer's Handbook*. 4th ed. Burbank, CA: BOMG Publishing, 2017.

Porges, Stephen W. *The Polyvagal Theory: Neurophysiological Foundations of Emotions, Attachment, Communication, and Self-Regulation*. 1st ed. The Norton Series on Interpersonal Neurobiology. New York: W. W. Norton, 2011.

Roads, Curtis. *Composing Electronic Music: A New Aesthetic*. Oxford; New York; Auckland; Cape Town: Oxford University Press, 2015.

Roginska, Agnieszka, and Paul Geluso, eds. *Immersive Sound: The Art and Science of Binaural and Multi-Channel Audio*. First published. Audio Engineering Society Presents. New York; London: Routledge, Taylor & Francis Group, 2018.

Ruiz, Sandra, and Hypatia Vourloumis. *Formless Formation: Vignettes for the End of This World*. Colchester: Minor Compositions, 2021.

Rumsey, Francis. *Spatial Audio*. Music Technology Series. Oxford; Boston, MA: Focal Press, 2001.

Shapiro, Livia. *The Somatic Therapy Workbook: Stress-Relieving Exercises for Strengthening the Mind-Body Connection and Sparking Emotional and Physical Healing*. Berkeley, CA: Ulysses Press, 2020.

Smalley, Denis. 'Spectromorphology: Explaining Sound-Shapes'. *Organised Sound* 2, no. 2 (August 1997): 107–26. https://doi.org/10.1017/S1355771897009059.

Tuan, Yi-fu. *Space and Place: The Perspective of Experience*. Minneapolis: University of Minnesota Press, 2002.

Wu, Jiayue Cecilia. 'From Physical to Spiritual: Defining the Practice of Embodied Sonic Meditation'. *Organised Sound* 25, no. 3 (2020): 307–20. https://doi.org/10.1017/S1355771820000266.

Zotter, Franz, and Matthias Frank. *Ambisonics: A Practical 3D Audio Theory for Recording, Studio Production, Sound Reinforcement, and Virtual Reality*. Cham, Switzerland: SpringerOpen, 2019.

Index

Note: *Italic* page numbers refer to figures.

Ableton 65, 161, 162, 171; Ableton Compressor 21
acoustics of space 34, 56
active equalisation 12
Adam Audio 57
adaptive equalisation: and AI 16; iZotope Neutron 16; Sonible 16
ADSR 95, *95*
advanced ear sensitivity 52–5, 85; amplitude recognition 53; compression, saturation, clipping 53; developing 108; ear training boot camp 52–5; frequency recognition 52–3; practice 52; space and phase 53
aesthetic identity 31, 121; optimising 6
AI-driven reverbs 34
AI tools 87, 90, 109, 152–3; adaptive processing for noise reduction 153; assistants 153; for emulation of analog processes 153; online full service for mixing/mastering 153
ambient sound 24, 106, 144
Ambi HD Decoder 166
ambisonic formats/mixing 157, 163–5, 167–9; encoder/panner 163; limitations 164; Reaper 164; routing 163, *164*; third-party ambisonics software 164
ambiX format 168
amplitude recognition 53–5
AMRAC Room mode Calculator 58, 60, 61
AMS 34
analog compressors 21–3
analog emulation plugins 73
analog gear 22, 23, 27, 48, 73, 151
analog metering scale *vs.* digital metering scale 40, *41*
analog meters 40
analog systems 40, 42, 73, 96
analytic listening 3, 76
APC 57

API 12, 49; API 550 12, 14, 15, *15*
Apple Music 155–7, 162
arrangement problems 99
artefacts 87, 96, 97, 102, 111; creating 27; mechanical 28–9; parallel processing and 103; phase 16, 32; saturation 26
artificial intelligence (AI) 152–4; and adaptive equalisation 16; as co-creative tool 153; copying and make music boring 153; into mixing process 152, 153; process 123; tools *see* AI tools
Art of Mixing, The (Gibson) 81
attack and release and knee 22; hard knee 22; soft knee 22
audio processing 2, 9, 10, 67, 118
auditory working memory 65

back bus technique 74
backing vocals (BV) 122–6; cleaning up 123; cohesion and blending 124; group *vs.* individual 125; optimising individual layers 123–4; panning concepts 124; pitching and timing 125; positioning relative to lead 124–5
bass 109–11, 133; relationship with kick 12, 63, 64, 89, 134
bass trapping 59–61
bedroom mixing 58–9; selecting monitors 59; setting up 59
BEDS 162, 168
bespoke listening environments 157, 159
bespoke max4live patches 171
bias 29, 64
binaural mixing 58, 165–6; HRTF 166; SOFA files 166
Bjork 156
Blumelin 157
bottom up mixing 76
Brauer, Michal 74

Brauerisation 74
Buchla 156

Camarq 11
channel-based audio 160–1; limitations 161; mix in 161; surround panner 160–1, *160*
channel EQ in Logic 13
cheat codes 86
chorus/phaser effects 32
classical music 80, 149
cleaning up 84–7, 118, 123
clicks and pops 87, 118, 123
clip gain automation 119
clipping 26, 39, 53, 151
colour coding scheme 68
comb filtering *see* phase
comparative analysis 63, 85
comparative listening 3, 64, 166
compression 18, 53, 94, 96, 100, 138, 143; attack and release and knee 22; compressor types 18–23, *19*; gain reduction and threshold 22–3; levelling/adjusting amplitude over time 18; makeup gain 23; as method for envelope shaping 105; multi-band 97, 110; parallel processing 120; threshold 22; two-stage 119–20; and vocal mixing 118–19
compressors 18–23, *19*, 100, 105, 114; characteristics 20; digital compressors 18, 21; FET compressors 18, 20–1; opto compressors 18, *19*; tube compressors 18–20; VCA compressors 18, 20, 21, 124, 146
continual resonances 117
controls 14–16; different settings 15–16; EQ slopes 14, *14*; frequencies adjustment 15; maximum gain adjustment 15
convolution reverbs 34, 171
critical bands 138
critical listening 1, 3–6, 25, 77, 79, 85–7, 142, 144; process 76; skills 31, 48, 62, 67, 76
Csikszentmihalyi, Mihaly 66; *Psychology of Optimal Experience, The* 66
cut boost technique 12

Daft Punk 25, 106
dance music 25, 80, 104, 106
dBFS 39, 40, 42
dBU, definition 40
decision-making 1, 5, 6, 89, 144
decoding 167, 169
deductive reasoning 11
de-essing 87, 97
delays 31–7, 102, 112, 114, 122; parallel 121–2; tape 122
depth 82, 113
'Designing Sound' (Farnell) 33

diagnostic listening 3
digital audio resolution 87
digital audio workstation (DAW) 13, 34, 38, 40, 144, 162, 165; channel EQ in Logic 13; EQ Eight in Ableton 13; standard meters in 38
digital clipping, definition 27
digital compressors 21; Ableton Compressor 21; Fab Filter 21; Logic Platinum Compressor 21; as one stop compressors 21
digital distortions 26, 39
digital equalisers 13, 16, 87; Fab Filter Pro Q 3 13; Waves Q10 13
digital reverbs 113; AMS 34; Lexicon 34; Ursa Major 34
digital technology 34, 156
discriminatory listening 3
distortion, definition 26; *see also specific types*
Dolby Atmos 72, 155–7, 159, 162–3, 165, 166, 169, 171; combination of two approaches 162; Dolby 7.1.4 165; Dolby BWF file 165, 168; Dolby Renderer program 162; limitations 162; mixing 155; routing in 162, *163*
downward expanding 96, 105–6
DSP power 35, 61, 144, 156
DTS 156
ducking 4, 25, 97, 106, 110
dynamic equalisation 97, 110, 111, 114, 118; *vs.* multiband compression 24
dynamic processing 24–6, 100, 105, 110–11; dynamic equalisation *vs.* multiband compression 24; gates and expanders 24; sidechaining 25; *see also* compressors
dynamics methods 118–20; clip gain automation 119; compressor selection 119; parallel processing compression 120; two-stage compression 119–20

ear training boot camp 52–5; advanced exercises 55; beginner 54; intermediate exercises 54–5
echoic memory 65, 75, 85
echo threshold 32; listening to Reich's early Phase Pieces 32; stereo imager/Haas effect *vs.* chorus/phaser 32–3
EDM Dubstep 82
Eight Channel 157–60, 162
electroacoustic music 143, 144
Elevado, Russell 175
embodiment: becoming embodied 69; and mixing 69; practising 69–70
Empirical Labs Distressor manual 23
encoding 167, 169
Envelope4Live 171

envelopes and shapes 95–6, 100, 110–11; compression as method for 105; for groove-based music 104–5; tools for adjusting 95–6
EQ Eight in Ableton 13
equalisers/equalisation (EQ) 10–12, *11*, 48, 49, 82, 100, 110, 124, 138, 149, 150; active 12; API 550 12, 14, 15, *15*; character 120, 124; digital 13, 87; dynamic 97; linear phase 117; Neve 1073 12–16, *15*, 120; passive 11–12, 150; and phase 16–17; slopes 14, *14*; wild curves 86
even-order harmonics 27
experimental music 115; experimental electronic music 80

Fab Filter 21; Fab Filter Pro Q 3 13
faders 53, 64, 82, 94, 124, 129
Fairchild 670 tube compressor 146
Farnell, Andy 33; 'Designing Sound' 33
Fibonacci series 129
field effects transistor (FET) compressors 20, 21; 1176 20–2, 119
first reflection path 59, *60*, 61
5.1 format *see* Six Channel
Fletcher Munson's curve 137, 138
Four Channel 158
14 Channel 159
frequency 84–7, 94; adjustment 11, 15; equalisation 86–7; optimising individual sounds 84; recognition 52–3, 87; solo/not solo 84–7; sweep/not sweep 85–6; using equaliser type 87
frequency changes 28; content degradation 28; low mid region 28; upper harmonic additions 28
frequency masking 74, 90–1, *91*, 99–100, 109, 138
frequency register and arrangement 88–90; adding bass instruments 89; adjusting 90; chord voicing 90; decision making 89; frequency range 88, 89; octave displacement 90
Freudian Psychoanalysis 68
Full Scale DB metering system 38

gain reduction and threshold 22–3
gain staging 73–4, 147
gates 96, 105–6; and expanders 24
genre and loudness 148–51; arrangements 149; and mastering for mixer 150–1; and mix bus/master 149–50; optimisation during mixing 149
Gibson, David 81; *Art of Mixing, The* 81
'glue' compressor *see* SSL, SSL G Bus
golden mean 129

groove-based music 104–6; compression and envelope 105; envelopes and envelop shaping 104–5; gating and downward expanding 105–6; getting into 104; psychoacoustic concept of streaming 106
groups of sounds 102; side effects 102

Haas effect 32–3, 94, 121, 122, 138
hard clipping 26, 50, 53
harmonic dissonances 27
harmonic distortion 26–8, 53, 90; definition 27
Harrison 11
headphone listening 156, 158, 166
headphone monitoring 57–8
Head Related Transfer Functions (HRTF) 138, 166
height 82
high-hat patterns 105–6
home cinema 156–7

immersive mixing/audio *see* spatial mixing
individual sounds 24, 33, 75, 84, 101, 106, 113–14
inductors 12
initial balances creation 81–2
insert processing: for source bonding 103
'It's Gonna Rain' (1969) 32
iTunes 42
iZotope 118; iZotope Neutron 16; iZotope RX 87, 153

jazz 6, 20, 115, 149

kick 99, 108–11, 133; connecting elements to each other through side chaining and envelop 105–6; relationship with bass 12, 63, 64, 89, 134
kick drums 18, 24, 85, 95, 106, 171

LA2A compressor 19, 22–3, 119, 121
law of the first wavefront *see* Haas effect
lead vocals/voice mixing 115, *121*; AI resonance tool for cleaning up sound 118; automating/throwing to delays/more complex effects 122; bedroom recording room mode 117; challenging for modern mixing 115–16; character equalisation 120; clearing unwanted frequencies 117; clicks, pops, mouth sounds 118; continual persistent resonances 117; dynamics methods 118–20; intermittent resonance 118; phase and space(s) 121; proximity effect 117; quality of 115, 116, 119; saturation and analog warmth 120–1; spaces in parallel 121–2; stacking 122; stereo imaging and doubler type plugins

122; working with parallel reverbs/delays 122
level balancing 11, 16–18, 42, 107
Lexicon 34
LFE channel *see* surround panner
linear function 26
listening, dimensions of 81
listening environment 56–61; making alterations to room 58–61; monitors/monitoring 57–8; room 58
LKFS *see* Loudness Units Full Scale (LUFS)
lo-fi analog unit 102
Logic 171; Logic Platinum Compressor 21; Logic Pro 157, 164
loudness 148–51; genre and *see* genre and loudness; matching 43, 64, 75, 151; wars 148, 170
Loudness Units Full Scale (LUFS) 42–3, 148; calibration 43; practical uses 43; reading 43; in streaming sites 42; using 42
low-end theory 107–11; adjusting frequency 109; challenges and solution 107–8; dynamics and envelope 110–11; phase issues 110; relationship between parts 108–9; saturation 111; side chaining 109–10
low-frequency absorption 59, 61
Low Frequency Environment (LFE) 168, 170

makeup gain 23
Manley Massive Passive 12
maximum gain adjustment 15
Melodyne 126, 127
metadata 161, 168
meters/metering 38, 72–3; analog 40; digital 38–9; peak 38–9; RMS 39; VU 40, *41*, 42, 73, *73*
mix aesthetics 140–1; *vs.* sonic aesthetic 140–1
mix bus processing 146–7; evolution 146; practice tips 147; setting up 147; top-down mixing 147
mix clarity 36, 64, 106
mix engineer in collaborative process 134–5; not changing but optimising 135; stages and focus 134–5
mixing: building momentum 128; composing joins 129; definition 1, 3; finishing 70–1; golden mean 129; individual signature style 140–1; performance and imperfection 128; subtle changes and fader automations 129; top-down 147; vocals in 63; *see also individual entries*
mixing and production 1, 2, 87, 116, 132–3; adding processing on top of processing 144; challenge and focus 143; combing 143; developing workflow 144; different genres and experimentalism 144; flatten and commit 144; orchestration and make adjustments to arrangement 143; preparation 132; putting off decisions 144; separating 142–3; tips for bouncing out stems/flattening tracks 132–3
mixing and psychoacoustics 136–9; acoustic reflex and dulling effect after exposure 136–7; compression 138; critical bands 138; directional illusion 138–9; equalisation 138; Haas effect 138; inter-aural time and intensity difference 138–9; loudness perception 137; noise exposure and hearing sensitivity 136; precedence effect 138; volume 137
mixing components and relationships 3, 3–6; aesthetic identity 6; critical listening 3–5; decision making 5; technical processing 5–6
mixing consoles 10–11, *11*, 146, 156
mixing toolkit 45–51; building 45–46; confidence of using 49; expanding/adjusting 49; expanding with analog approach 49; learning 48; quiz to improve mixing 50–1; thirty essential tools 46–7
mix picture building 81
monitors/monitoring 57–8; headphone 57–8; stick with your 57
mono listening *see* One Channel
mouth de-noiser 87
mouth sounds 118
muddiness 28, 34, 36, 64, 88, 91, 110
multi-purpose space, treatment in 59–60; absorption *vs.* diffusion 59, 61; bass trapping 59–60; ceiling and floor 60; first reflection path 59, *60*; speaker boundary interference 59
multi-speaker listening 58, 156
muting 96

neo-soul sounds 175
Neve 11, 12; channel strips 13; Neve 1073 12–16, *15*, 49, 120; Neve 2024 13; Neve 8028 146; Neve 33609 146
NoiseMakers ambiVerb 171
non-linear function, definition 26
non-linear waveshaping 27, 29

object-based formats 58, 161, 165; limitations 162; mix in 162; object panner 161
octaphonic *see* Eight Channel
old school delays 35
One Channel 158, 169
opto compressors 19; LA2A 19, 22, 23, 119, 121
organised sound and prioritising 80–2
overdrive 27

panning 53, 82, 91, 110, 124, 138, 166–7, 169; amplitude 160; feather 124; LCR 124; random 124
parallel compression 74, 120
parallel processing 122, 171; compression 120; for less artefact 103
parallel reverbs/delays 122
passive equalisation 11–12, 150; Manley Massive Passive 12; Pultec EQP1A 12
peak meters 38, 39; measurement scale of 39; measuring 38–9; practical uses 39; usefulness 39
Pensado, Dave 69
perception: and depth 36, 64, 113, 114; and intimacy 36; of width 113
phase 30–3; cancellation 86, 110; correlation meters 110; definition 30; distortion 17, 26, 99; echo threshold 32; and equalisation 16–17; in/out of phase 30, *30*; issues 30–1, 98, 110; magic 31; processes 31, 82, 124, 138; psychoacoustics and precedence effect 31; space and 53
pitching and timing 125
plugins 10, 34, 35, 45, 46, 122, 171
Power, Bob 175
precedence effect 31–2, 138
pre-mix preparation 72–7; DAW/desk set up 72; gain staging 73–4; making own choice and developing own workflow 72; mixing preparations 77; mixing training 77; routing 74–5, *75*; sample rate and bit depth 72; setting up meters 72–3; setting up mix bus 75–6; setting up references 75; workflow direction 76
Pro Tools 157, 165
psychoacoustic concept of streaming 106; of rhythmic elements 106
psychoacoustic(s) 31–2, 36, 56, 81, 82, 111, 113, 136–9; anomalies 57; definition 136; effect 53; related to mixing 136–9; *see also* mixing and psychoacoustics
Psychology of Optimal Experience, The (Csikszentmihalyi) 66
Pultec EQP1A 12, 49, 91

quadraphonic mixing 155, 156, 158

Reaper 164, 169
references 62–5; bypass mix processing 65; mix engineers using 62; and personalised reference playlist 62, 63; setting up 64–5, 75; tool for mix engineers 62; uses 63–4, 108; working less with 141
Reich, Steve: early Phase Pieces 32
rendering 167–9; bouncing 167–8; monitoring 167

reverbs 34–7, 112, 114; AI-driven 34; ambisonic 171; convolution 34, 171; digital 34, 113; long 122; matching with style 113; multimono 171; parallel 122; send 124; types 34, 102
REW Wizard 58, 60
ringing resonance 3, 12, 17, 50
ripple effect 87
RMS meters 39
Rockit 57
room analysis tools 58
Room EQ Wizard 61
Root Means Squared (RMS) meters 39; measuring scale 40; usefulness 40
routing 74–5, *75*, 143, 162, *163*, 163, *164*, 168–70

saturation 35, 53, 96, 100, 111, 150; and analog warmth 120–1; to create thickness 101–2
saturation and distortion 26–9; creating artefacts 27; definitions 26–7; frequency changes 28; mechanical artefacts 28–9; smearing transient 28
Scheps, Andrew 74, 120
2nd-order harmonic distortion (THD) 27, 101; definition 27
self-treatment in mix studio 60–1; room modes and anti-modes 60–1
7.1 format *see* Eight Channel
Shepards tone 82–3
side-chain compression 106; side-chain gating 106
side chaining 25, 92, 97, 105–6, 109–10, 114, 143
Six Channel 155, 157, 158, 164
SOFA files 165
soft clipping 27, 50, 53
Sonarworks 58, 60, 61
Sonible 16
sonic aesthetic 2, 34, 86; *vs.* mix aesthetics 140–1
Soothe 118
Soundtoys Microshift 124
source bonding 90, 103, 113
source positioning 171; mono *vs.* stereo localisation diffusion 171–2, *172*
space: aligning with aesthetic 112; to create thickness 102–3; and emotion 113; perceptions of 83
space processes/processing 34–44, 138; adjusting and tuning 36; analog meters 40; analog *vs.* algorithmic *vs.* convolution/impulse response *vs.* AI 34; connecting to aesthetic of production 35; dBU 40; digital meters 38–9; focus on fixing and

cleaning first 113; levels of mixing 40; LUFS 42–3; old school delays 35; parallel/inserting delays 35–6; parallel tracks for cohesion and clarity 114; peak meters 38; perception of body/size of individual sounds 113–14; perspectives and uses 35; reverb matching with style 113; RMS meters 39–40; time-based processes and 34, 82, 83; tuning 114; using many reverbs and delays 36–7; working together 114

space repositioning 91–2; phase/space 92; shape and dynamics 91; simple balances 91

SPARTA 164, 169

Spat for Ableton packages 171

spatial audio 58, 155–8

spatialising 167, 169

spatial mixing 155–73; 14 Channel 159; 20+ channels 159; approaches to spatial mixing 160–5; binaural mixing/monitoring 165–6; definition 155; depth and space 171; Eight Channel 160; encoding/decoding 167; Four Channel 158; frequency and dynamic processing 170; loudness 170; Low Frequency Environment (LFE) 168; low frequency management 170; monitoring 165; movement *vs.* stability 172; One Channel 158; panning 166–7; prep audio 169; rendering 167; routing/bussing 168; setting up and routing 169; Six Channel 158; source positioning 171; spatial audio by Apple Music 155–6; spatial audio history 156–8; spatialising 167; stereo and 168–9; Two Channel 158; workflow 172–3

Spector, Phil 146

Spotify 42, 43

SSL 11, 12, 49; channel strips 13; E series 15; 4000 series 146; G series 15, 16; SSL 611 12, 14; SSL G Bus 20, 146

"Standards for 'Safe Listening': Past, Present, and Future" (Kawamori, Best and Laureyns) 136

stereo field 53, 82, 91, 94, 124

stereo format 155–8; mid/side components in 157

stereo imaging/imager 31, 32, 82, 122, 124, 138

stereo miking techniques 34, 163

stereo mixing 53, 81, 82, 138, 169

stereophonic listening 139

sub-bass 60, 107, 109–11

surround panner 160–1, *160*

tape delays/loops 35, 122

tape hiss 27, 28

tape machines 27, 28, 32

tape saturation 27, 100–2, 111

technical processing 3, 5–6, 135, 174

techno 6, 104, 143, 149

thickness: saturation to create 101–2; space to create 102–3

third-party ambisonics software 164, 169

1176 compressor 20–2, 119

3D field 161, 163, 168

3D sound mix 166

3rd-order harmonic distortion (THD) 27, 101; definition 27

360° mixing *see* spatial mixing

timbre 26, 35, 48

time processes 34, 100, 102, 105, 113

tonal balance 96, 114

tonal shaping 100, 120

top down mixing 76

transformers 11, 13, *14*, 27

transient designer 105

transients 63–4, 94; controlling/reducing 94; diffusing 94; stereo field 94; time and space continuum 94

transient shaper 96

transistors 11, 20

tube compressors 19–20, 23; Tube-Tech 19–21

tube saturation 27, 100, 101

Tube-Tech compressor 19–21; Tube-Tech CL1B 23

Two Channel 158

two-stage compression 119–20

unity, concept of 73

upper harmonic additions 28

Ursa Major 34

VCA compressors 18, 20, 21, 124, 146

visual representation of sounds 95, *95*

vocal harmonies 75, 95, 124

Voltage Controlled Amplifier (VCA) compression 75; SSL G Bus 20

Volume Units (VU) meters 40, *41*, 73, *73*; correlation with other meters *41*; measuring 40, 42; relevance 42

Voodoo (D'Angelo) 175

VR 157

VU meter 23, 73

wave field synthesis 159

waveshaping 26; definition 26; types 53

Waves Q10 13

width 82, 113

workflow/flow states 66–71; achieving 68; distractions 67–8; embodiment 68–70; establishing 70–1; as goal-directed rule-bound action system 67; skill acquisition 67

wow and flutter 28

XR 157

Made in the USA
Monee, IL
03 May 2026